Control...
Arduino/Genuino

by

David Leithauser

Copyright © 2018 David Leithauser

All Rights Reserved

Table of Contents

Introduction 3
Turning Motors On and Off with Relays 5
Reversing DC Motors with Relays 13
Speed Control for DC Motors with Transistors 17
Reversible with Speed Control 27
Controlling DC Motors with Motor Controllers 30
Unipolar Stepper Motors 39
Bipolar Stepper Motors 61
Servo Motors 73
Brushless Motors with ESC 82
Additional Information 98

Introduction

This book, as the title indicates, is about controlling motors using Arduinos (called Genuinos outside the United States). Although the examples in this book are about motors, some of the techniques discussed can be used to control other devices. For example, the discussions of turning motors on and off and controlling their speed could also be applied to other devices like lights.

I start out with simple direct current (DC) motors, covering turning them on and off, controlling their speed, and reversing direction. I then get into more complicated motors. I discuss stepper motors, a type of precision motor that can be used for detailed control in devices like 3D printers, CNC routers, robot arms, etc. I also discuss servo motors, which as useful for setting positions, and brushless motors, which are good for high speeds like you might need for drone or model plane propellers, fans, model boat propellers, high speed model cars, and any other application that requires high speeds.

This book covers both electronics and programming, so there will be electronic schematics and Arduino sketches. I assume that the reader is already familiar with the basics of Arduinos, such as how to install the IDE and load sketches, so I do not go into details on topics like that. I use Arduino Uno as the basis for designs in this book, but other models will work for most of the applications discussed in this book.

The code in this book is available under the MIT license. This license gives you the right to use and modify the code in any way you like. Under this license, you agree not to hold the author liable for any damage caused by the use of misuse of the sketches. Likewise, you may use and modify the material in this book for any purpose provided you agree not to hold the author liable for any use of misuse of the information, schematics, or sketches described in this book.

Chapter 1

Turning Motors On and Off with Relays

This chapter will cover the very simplest task with motors, turning them on and off. In this chapter, I will discuss using relays. There are other ways to turn motors on and off, and some of these will be covered in later chapters. However, relays have several advantages as controllers. They can be with both AC (alternating current) or DC (direct current) motors, and most Arduino relays can be used on high voltages, such as 120 or even 220. There is also no problem with power loss or heating of the relays the way there can be in solid state devices like transistors. The descriptions in this chapter, incidentally, can be used on other devices besides motors, such as lights, heaters, etc.

A relay is a device that allows a low voltage and small current to switch on and off a larger voltage and/or current. The small voltage and current are applied to the coil, and this turns on or off a switch. There are various characteristics of relays.

The most significant characteristic of a relay is the number of throws it has. The throw is the number of positions the relay switch can be in. Possible positions are open and closed. A single throw relay normally has a normally open switch, which means that no current can flow through the switch if there is no power to the relay coil. It is possible but extremely rare to have a single throw relay that is normally closed, meaning that power can flow through the switch unless power is applied to the relay coil. Most relays, however, are double throw relays. This means that the switch has both normally open and normally closed positions. When there is no power to the relay coil, current can flow through the normally closed connections but not the normally open connections. Figure 1.1 shows the symbol for a double throw relay.

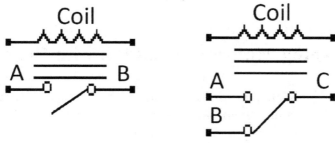

Figure 1.1

The symbol on the left is the single throw relay. When no power is applied to the coil, no current can flow from A to B. When power is applied, current can flow. The symbol on the right is a double throw relay. Current can flow from B to C when there is no power to the coil, and current can flow from A to C when power is applied to the coil. Because

Relays can also have more than one switch, called a pole. For example, a relay with one switch is called a single pole and a relay with two switches is called a double pole. One fairly common type of relay is a double pole, double throw (DPDT) relay. The symbol for this is shown in Figure 1.2.

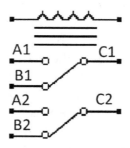

Figure 1.2

In a DPDT relay, power can flow from B1 to C1 and B2 to C2 when there is no power to the coil, and from A1 to C1 and A2 to C2 when there is power to the coil.

One small problem with using relays with Arduino is that the digital output pins that you will use to turn the relay on and off do not have enough power to energize the relay. That is, they cannot provide enough current to the coil for the coil to flip the switch. Because of this, relays designed for Arduinos have a pin to be connected to the 5V pin on the Arduino to actually power the relay coil. The digital output from the Arduino then simply tells the relay module to turn on the relay, rather than actually powering the relay. Figure 1.3 shows two Arduino relay modules.

Figure 1.3

The module on the left has two relays, the module on the right has one. You can see the three connection pins on the module on the right. The pin on the right should be connected to one of the digital output pins on Arduino. The center pin should be connected to the Arduino 5V pin, and the one on the right to any one of the Arduino GND pins. Notice that the module on the left has four pins (on the middle of the left side in the picture). The top one should be connected to the Arduino 5V pin, the second one to one of the Arduino GND pins, and the bottom two each go to one of the Arduino digital output pins. Each of the Arduino digital pins controls one of the relays.

Different manufacturers use different labels for the pins. The pin that gets connected to the Arduino 5V may be labeled +, V, or VCC. The ground connection may be labeled -, G, or GND. The digital input may be labeled S (for signal) or various variations of IN (for input). If the module has several relays, the IN pins will be numbered.

The schematic for using a relay to control a motor is shown in Figure 1.4.

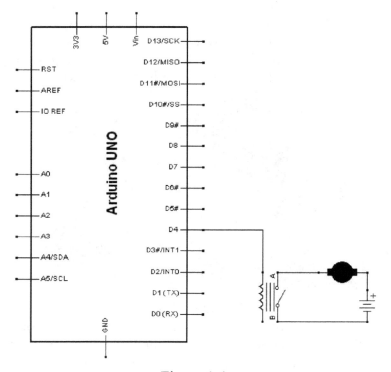

Figure 1.4

In this diagram, I have connected the relay to digital output 4, but you can use whichever one you want. Also, I have used a single pole, single throw (SPST) relay, the minimum required for this operation. It is worth noting that you can always use a more complex relay than the one specified in any schematic. For example, you can use a single pole, double throw (SPDT) relay or a double pole single throw (DPST) relay or a double pole double throw relay for any project that needs a SPST relay. You simply do not use the extra connections.

Although it is normal in schematics to show the other coil contact connection, I will not do that in this book for several reasons. For one thing, some relays turn on when the input from the Arduino digital output goes positive, and some turn on when the digital output goes negative. In the latter case, it would be incorrect and confusing to show the other coil connection connected to ground (GND) when it should be connected to positive if digital output going to ground turns the relay one. Also, as explained above, for Arduino relay modules you are really connecting only the digital output of the Arduino to each relay, and the connections to ground and power (5V) are handled internally within the module. Therefore, in this book, I will show only the digital connection to the relay.

The sketch for turning the relay on and off is extremely simple too. Of course, you will decide what input you want to control the relay, and therefore the motor. In this book, I will use the serial input to demonstrate in most sketches for simplicity. A simple example is shown in Sketch 1.1. In this sketch, you turn on the relay by typing ON in the serial monitor text box and pressing ENTER or clicking on the Send button. You turn off the relay by typing OFF in the serial monitor text box and pressing ENTER or clicking on the Send button.

```
#define relayOn HIGH
#define relayOff LOW
#define relayPin 4
String relayCommand;

void setup() {
  // put your setup code here, to run once:
  pinMode(relayPin, OUTPUT);
  digitalWrite(relayPin, relayOff);
  Serial.begin(9600);
} // End of setup
```

```
void loop() {
  // put your main code here, to run repeatedly:
  if (Serial.available() > 0) {
    relayCommand = Serial.readString(); //get string from serial buffer
    relayCommand.toUpperCase();
    Serial.println (relayCommand);
    if (relayCommand == "ON") {digitalWrite(relayPin, relayOn);}
    if (relayCommand == "OFF") {digitalWrite(relayPin, relayOff);}
  } // End of if statement
} //End loop
```
Sketch 1.1

This sketch starts by using the define statement to set certain values that you might want to change later. For example, it sets relayOn to HIGH and relayOff to LOW. The reason for this is that, as mentioned previously, some relays are turned on by applying +5 volts to the signal input and some by applying ground to this. By defining relayOn and relayOff at the beginning and then using these defined values through the sketch, you can change all of these at once throughout the sketch by just changing the define statements, rather than going through the entire sketch and changing every HIGH to LOW and every LOW to HIGH if you change the type of relay you are using. It certainly makes it easier on you, the reader, if you can just change these two lines in the sketch if you decide to use a ground triggered relay instead of the positive triggered relay I used. Likewise, I defined the relay pin to 4 at the beginning, so that you can change this easily, rather than putting 4 in the changing every digitalWrite statement and forcing you to find and change every one of these in the code if you decide to use a different pin. The variable relayCommand will hold the string that you send from the serial monitor.

In the setup routine, I first set the pinMode of relayPin to OUTPUT. I then use digitalWrite to set the output of this pin to the off value for the relay to make sure the relay starts out turned off when the sketch first runs. This is important if you are using a relay that turns on when the input is ground, because in some models of Arduino pinMode automatically sets the pin HIGH by default, so the relay would start out turn on as soon as you turn on the Arduino. The last command in the setup routine is to initialize serial communications so you will be able to send commands to the Arduino from the serial monitor.

In the loop routine, the line
if (Serial.available()) {
checks to see if any characters have come in from the serial monitor. If any have, the
relayCommand = Serial.readString();
line reads the string of characters from the serial monitor and stores them in the string variable relayCommand. The Serial.print line sends the string back to the monitor. This is just for error testing, so that if the relay does not respond properly you can see if you just mistyped ON or OFF. The following two if statements then test to see which of the two commands you typed. The
if (relayCommand == "ON") {digitalWrite(relayPin, relayOn);}
statement tests to see if the command is ON, and if so sets the relay pin to the on state. The
if (relayCommand == "OFF") {digitalWrite(relayPin, relayOff);}
statement turns the relay off if you typed OFF. Important: Note that in if statements, you need a double equal sign (two equal signs in a row) to indicate that you are testing whether they are equal. If you only use one equal sign, the statement would assign "ON" or "OFF" to relayCommand instead of testing to see whether it is already equal.

After testing to see if the command was ON or OFF, the program flow exits the
if (Serial.available()) {

test and then goes back through the loop. Note that if there is no message waiting from the serial monitor, none of the code is executed. All of the code is within the
if (Serial.available()) {
condition.

      Of course, in your application, you will almost certainly want to have the motor controlled by something more sophisticated than text command from the serial port. However, this chapter walks you through the basics of using relays, both from a hardware perspective and the basics of using the digitalWrite command to control relays. In the next chapter, I will go into making a basic direct current motor not only start and stop but reverse direction under Arduino control.

Chapter 2

Reversing DC Motors with Relays

For many application, you will want to reverse the motors. For example, a toy car or robot needs to be able to back up. In this chapter, I will give you schematics and code for this. This will require using the double pole double throw (DPDT) relay described in Chapter 1 to reverse the motor polarity. It will also use a SPST relay for turning the motor on and off, as in Chapter 1. The schematic is shown in Figure 2.1.

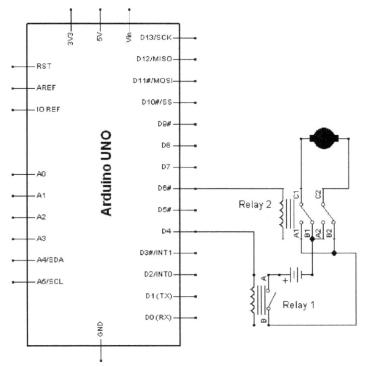

Figure 2.1

Each of the two wires to the motor should be connected to one of the two common terminals of the relay. The normally open contact on one switch must be connected to the normally closed contact of the other switch and also the power supply. The remaining normally open contact connected to the remaining normally closed contact, and to one contact of the SPST relay switch. The other switch contact of the SPST relay should be connected to the other wire of the power supply. In this way, the SPST relay turns the power off entirely when it is turned off, and the DPDT relay reverses the connection to the motor.

Each of the relay coils should be connected to one digital pin of the Arduino. It does not really matter for this project which ones. I have used 4 for the SPST relay and 6 for the DTDT relay.

Sketch 2.1 shows how to control this circuit. Again, it uses commands from the serial port as an example, but you can use any input to control it.

```
#define relayOn HIGH
#define relayOff LOW
#define onOffRelayPin 4
#define reverseRelayPin 6
String relayCommand;

void setup() {
  // put your setup code here, to run once:
  pinMode(onOffRelayPin, OUTPUT);
  digitalWrite(onOffRelayPin, relayOff);
  pinMode(reverseRelayPin, OUTPUT);
  digitalWrite(reverseRelayPin, relayOff);
  Serial.begin(9600);
} // End of setup

void loop() {
  // put your main code here, to run repeatedly:
  if (Serial.available()>0) {
```

```
    relayCommand = Serial.readString(); //gets one byte
from serial buffer
    relayCommand.toUpperCase();
    Serial.println (relayCommand);
    if (relayCommand == "ON")
{digitalWrite(onOffRelayPin, relayOn);}
    if (relayCommand == "OFF")
{digitalWrite(onOffRelayPin, relayOff);}
    if (relayCommand == "BACK")
{digitalWrite(reverseRelayPin, relayOn);}
    if (relayCommand == "AHEAD")
{digitalWrite(reverseRelayPin, relayOff);}
    } // End of if statement
} //End loop
```
Sketch 2.1

This is very similar to sketch 1.1, with a few lines added. I will discuss only the added lines.

First, I defined reverseRelayPin as pin 6 to set which pin would operate the reverse polarity relay, and changed relayPin to onOffRelayPin to clarify which relay that pin controlled. In the setup routine, I initialized this pin as an output pin using pinMode and set its initial output value to relayOff using digitalWrite. In the loop routine I added the lines

```
 if (relayCommand == "BACK")
{digitalWrite(reverseRelayPin, relayOn);}
 if (relayCommand == "AHEAD")
{digitalWrite(reverseRelayPin, relayOff);}
```

so that the text commands "BACK" and "AHEAD" will control the relay that controls whether the motors go forward (AHEAD) or in reverse (BACK). You could, of course, use any words you want for the commands and probably will not be using typed commands at all. The purpose of these sketches that use text commands is to demonstrate the principles explained in this book, and also give you some simple sketches to allow you to easily test whether you have the wiring right when you build the circuits.

Chapter 3

Speed Control for DC Motors with Transistors

Having the motors go backward and forward is important, but in many applications, such as robots, model cars, and drones, you will also want to control the speed of the motors. Fortunately, Arduinos have the ability to control the output of some of the digital outputs with pulse width modulation (PWM). This means that instead of just being on or off, the digital output can rapidly turn on and off hundreds or even thousands of times per second. You can control what percentage of the time the output is high and low. You can therefore control what percentage of the time an object, like a motor, has power. This can be used to control the speed of a motor, the brightness of a light, and so on.

Not all of the digital output are capable of PWM. On the Arduino Uno, it is pins 3, 5, 6, 9, 10, and 11. The pins capable of PWM are usually marked with a # symbol next to their number on the Arduino.

You control the percentage of the time a PWM digital pin is high (5 volts) using the command analogWrite. This command is similar to digitalWrite. However, digitalWrite can only write either LOW or HIGH. The command analogWrite can accept a number from 0 to 255, where 0 causes the output to always be low and 255 causes it to always be high. Numbers in between these two values cause the output to be high the ratio of the time that the value has to 255. For example, analogWrite(outputPin, 127) would cause the output pin to be high 50% of the time, because 127 is 50% of 255. Likewise, analogWrite(outputPin, 51) would cause outputPin to be high 20% of the time, since 51 is 20% of 255, and so on. Note that this is not a true analog output where the output voltage ranges from 0 to 5 volts. Instead, it is always either 0 or 5 volts, but the amount of time it is each varies.

Sketch 3.1 demonstrates how to use this by allowing you to input a number from 0 to 100 and having pin 5 be high this percentage of the time. Of course, using 0 stops the motor completely.

```
#define analogOutPin 5
String motorCommand;
int speedVar;

void setup() {
// put your setup code here, to run once:
  pinMode(analogOutPin, OUTPUT);
  analogWrite(analogOutPin, 0);
  Serial.begin(9600);
} // End of setup

void loop() {
  // put your main code here, to run repeatedly:
   if (Serial.available() > 0) {
     motorCommand = Serial.readString(); //get string from serial buffer
     Serial.print (motorCommand);
     speedVar = motorCommand.toInt();
     speedVar = map(speedVar, 0,100,0,255);
     Serial.print(" ");
     Serial.println(speedVar);
     analogWrite(analogOutPin, speedVar);
     motorCommand = "";
   } // End of if statement
} //End loop
```
                    Sketch 3.1

Since this is similar to Sketch 1.1, I will discuss only the differences. I have eliminated the relayOn and relayOff definitions, since there are no relays here. I have replaced the definition relayPin with analogOutPin for the same reason. Note that I changed the output pin from 4 to 3, because 4 is not one of the pins that has PWM capability. I also set up the integer variable speedVar, which will hold the speed you want to set the motor to.

Note that in the setup routine, I still use
pinMode(analogOutPin, OUTPUT);
The pin mode is still output, regardless of whether it is straight digital or PWM. I then set the initial output to 0, meaning always off, using
analogWrite(analogOutPin, 0);

In the loop routine, I have removed the line
relayCommand.toUpperCase();
because the input will be numbers, not letters, so there is no need to make sure they are capitalized. I then use the toInt() function to convert the string motorCommand into an integer number in the line
speedVar = motorCommand.toInt();
This converts motorCommand to an integer and stores it in speedVar. Note that speedVar will then equal a number from 0 to 100. Since the value you use for analogWrite should be from 0 to 255 instead of 0 to 100, I use the map function to map the 0 to 100 into 0 to 255 in the line
speedVar = map(speedVar,0,100,0,255);
This takes the input value of speedVar and converts it to a proportional number from 0 to 255, then stores it back in speedVar. The line
analogWrite(analogOutPin, speedVar);
then uses this number from 0 to 255 to control how much of the time the output of the pin designated as analogOutPin is 5 volts and how much of the time it is 0. The Serial.print statements as just for diagnostic purposes. Figure 3.1 shows an oscilloscope display of the output of two PWM pins. The top one is set for 20% high, and the bottom one for 80% high.

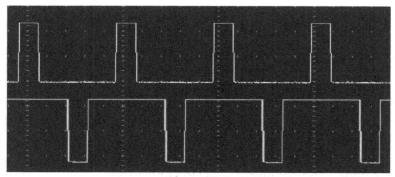

Figure 3.1

Now that we have a pin that we can control the output of more precisely, we need the hardware to control a motor from that pin. A relay, such as we used in the previous chapters, will not work. Remember that the output is switching hundreds or even thousands of times per second. The switch on a relay cannot turn on and off that fast. You need a solid state device like a transistor.

Transistors simply have an input connection where you apply a control voltage and another connection where you connect a load to. Then there is the ground connection, which applies to both the input and the load. There are various types of transistors. The most useful for us might be the bipolar junction transistor (BJT) and the metal–oxide–semiconductor field-effect transistor (MOSFET). The most useful type for connecting to the Arduino will be the MOSFET, because these have very high resistance at the gate, which you will be connecting the Arduino output to, so they draw almost no current. They also have very little resistance at the drain when they are conducting, so they make good switches. We will be using transistors only as switches with the Arduino, since the output of the Arduino is digital.

Figure 3.2 shows a MOSFET connected to an Arduino.

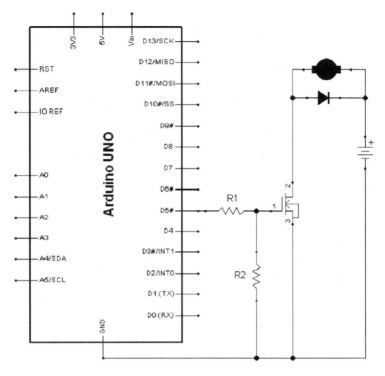

Figure 3.2

I have arbitrarily connected the transistor to output D5, but it could be connected to any of the digital outputs that have PWM capability. The R1 is just a safety to make sure that too much current is not drawn from the Arduino in case of a short or failure of the MOSFET, but normally there will be almost no current to the MOSFET gate and no voltage drop across the resistor. At typical value might be something like 220 ohms. R2 has a more important role. The gate of a MOSFET actually has some capacitance, so it can actually hold a charge. If you apply a positive voltage to the gate and then disconnect, the gate can actually hold the charge for a long time, causing the MOSFET to remain on. Normally this is not a problem with the Arduino, because the output is either HIGH or LOW, and when it goes LOW it will discharge the MOSFET gate. However, if power to the Arduino is turned off while the output is

HIGH, the gate can remain charged and the MOSFET continue to conduct. In addition, the gate is so sensitive that even radio waves or static electricity can charge the gate if the Arduino is disconnected or turned off. I recommend including R2 so that the gate is pulled to ground if no ongoing positive voltage is applied to the gate, to make OFF the default state for the MOSFET. A fairly high resistance, like 100 K, is sufficient to discharge the MOSFET and will not significantly affect the circuit. The actual values are not really that important, so long as R2 is many times as high as R1, so that you do not have a significant voltage drop across R1.

You may have noticed the diode across the motor. This is very important when using motors. Motors have magnetic coils that store energy. When the power is cut off, which happens many times per second in this design, this energy can be released as an electrical discharge. This discharge is often called inductive kickback or inductive flyback of the motor. The voltage of this discharge can be many times the original voltage applied to the motor. This discharge would try to go through the transistor, and could burn it out. The diode discharges this energy harmlessly. Note that the diode must be pointed in the opposite direction of the normal current flow, Otherwise, it would simply short out the motor and the motor would not move.

There are various MOSFETs you can use. One popular model for Arduino circuits is the IRF540, which you can buy on eBay for about $0.55 each when you buy them in batches of 10 if you do not mind waiting for a shipment from China. Figure 3.3 shows what the IRF540 looks like. In fact, many MOSFETs look like this. It is a common configuration. The numbers on the pins correspond to the numbers on the transistor shown in Figure 3.2, where 1 should be connected to the Arduino output, 2 to the motor and 3 to the common ground.

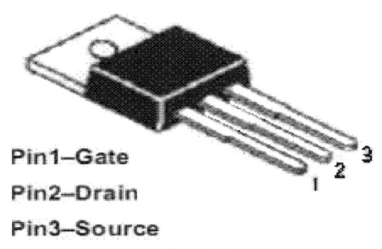

Pin1—Gate

Pin2—Drain

Pin3—Source

Figure 3.3

The IRF540 has a maximum drain to source voltage of 100 V, a maximum current of 33 amps at room temperature (maximum current allowed drops to 23 amps as the temperature rises to 100 C in case you plan on operating it in boiling water), and takes between 2 and 4 volts at the gate to cause it to pass a current (called the threshold voltage). The metal backing is a heat sink, and you can bolt it to a larger heat sink. If you need to use a voltage greater than 100 V in your circuit, there is the 2SK3568, which has a maximum drain to source voltage of 500 V, a maximum current of 12 amps, and also threshold voltage of 2 to 4 volts. There are many more types of MOSFET. The only real requirement of the MOSFET is that it has a threshold voltage less than 5 V (preferably closer to 3 V) so that the Arduino digital output of 5 V can trigger it. I do prefer to use N channel MOSFETs, which are the ones that the circuits in this book are designed for, because the voltage from the Arduino goes to ground, not the load.

If you need to draw more current than your MOSFET can handle, you can actually connect more MOSFETs in parallel. Just connect the gates of all the MOSFETs together, the drains of all the MOSFETs together, and the sources of all the MOSFETs together. This is shown in Figure 3.4.

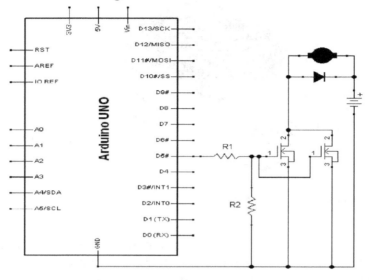

Figure 3.4

You can use BJT transistors instead of MOSFET if you want to. I recommend using NPN transistors if you do, because the load goes on the collector and does not interfere with the base to emitter current flow. An Arduino circuit using a BJT is shown in Figure 3.5. Note that the Arduino is connected through a resistor to the base of the transistor, labelled B. The emitter, labelled E, goes to the ground. The collector, labeled C, goes to the motor.

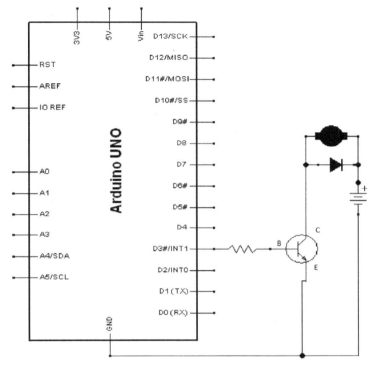

Figure 3.5

R1 is a current limiting resistor, to prevent the transistor from drawing too much current. The exact value you use can depend somewhat on which transistor you use and the exact value is not generally important. You want to use the highest resistance that will pass enough current to saturate the transistor base and allow the transistor to conduct enough current for your load. Generally, a 2.2K resistor is about right. If that is not allowing enough current to your load, reduce it to about 1 K. Note that no resistor is needed when using BJT because, unlike MOSFET, there is low input resistance and current can flow into the base, thus preventing a charge buildup.

As for what type of transistor to use, the TIP31C is a usually a good one. It can conduct 3 amps and can handle up to 100 V from the collector to the emitter, meaning that it can operate up to 100 V loads. Figure 3.6 shows a typical TIP31C transistor pinout.

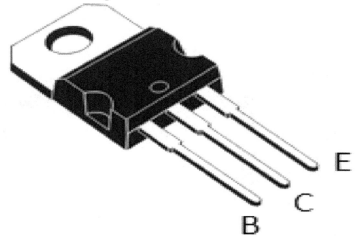

Figure 3.6

If you need to conduct more current, you can string BJT in parallel in parallel, as shown in Figure 3.7.

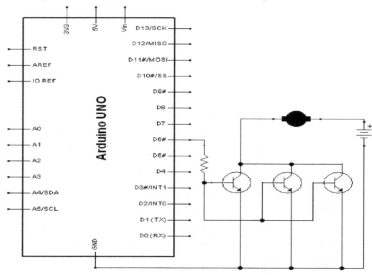

Figure 3.7

Chapter 4

Reversible with Speed Control

This is rather trivial at this point, but for the sake of thoroughness, I will combine the information in chapters 2 and 3 to create a sketch and schematics to enable you to both reverse and speed control a DC motor. The electronic schematic for this is shown in Figure 4.1.

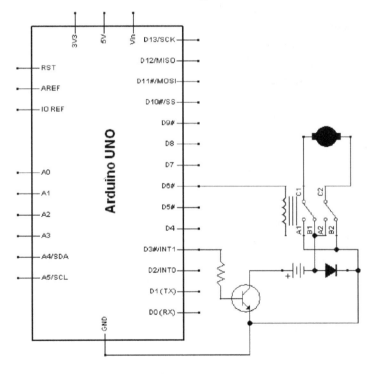

Figure 4.1

This is almost exactly the same as Figure 2.1, except the on-off relay is replaced by a transistor controlled by a PWM pin, allowing speed control instead of just turning the motor on and off.

The sketch for this circuit is Sketch 4.1, which is a combination of Sketch 2.1 and Sketch 3.1.

```
#define relayOn HIGH
#define relayOff LOW
#define analogOutPin 3
#define reverseRelayPin 8
String motorCommand;
int speedVar;

void setup() {
  // put your setup code here, to run once:
  pinMode(analogOutPin, OUTPUT);
  digitalWrite(analogOutPin, 0);
  pinMode(reverseRelayPin, OUTPUT);
  digitalWrite(reverseRelayPin, relayOff);
  Serial.begin(9600);
} // End of setup

void loop() {
  // put your main code here, to run repeatedly:
  if (Serial.available() > 0) {
    motorCommand = Serial.readString();
    motorCommand.toUpperCase();
    Serial.println (motorCommand);
    if (motorCommand == "BACK")
{digitalWrite(reverseRelayPin, relayOn);}
    if (motorCommand == "AHEAD")
{digitalWrite(reverseRelayPin, relayOff);}
    if (motorCommand >= "0" && motorCommand <= "9") {
      speedVar = motorCommand.toInt();
      speedVar = map(speedVar, 0,100,0,255);
      Serial.print("Speed = ");
      Serial.println(speedVar);
      analogWrite(analogOutPin, speedVar);
    }
  } // End of if statement
```

} //End loop

Sketch 4.1

In this I have changed the name analogOutPin pin number from 5 to 3 because pin 3 has PWM capability. I changed the pin number of reverseRelayPin from 6 to 8, but that was just to make more room on the schematic for the transistor. I then replaced all the code in Sketch 2.1 that controls the on-off relay with the code from Sketch 3.1 that controls speed. Note the line
 if (motorCommand >= "0" && motorCommand <= "9") {
This checks to see if the first character in motorCommand is a number. If it is not, the if clause is skipped. Without this, the toInt() function would read anything that is not a number as 0, and stop the motor.

Chapter 5

Controlling DC Motors with Motor Controllers

So far, I have talked about controlling motors using very basic components, relays and transistors. There are, however, complete devices designed to provide control of motors. Many of these are basically enhancements of the transistor circuits described in Chapters 3 and 4. They have input pins that accept signals from Arduino output pins and output the same signal to output connections with more current or higher voltages than they input. In this chapter, I will discuss using the L298N motor controller to control DC motors.

In my opinions, the L298N motor controller has some disadvantages over the techniques discussed in previous chapters. For one thing, it requires three outputs from the Arduino, instead of the one (for on/off or speed control only) or two (for reversibility plus on/off or speed control). That can cut down on available output pins from your Arduino. I will, however, discuss ways to use just ground and 5V outputs in some cases where you only want limited functionality, such as speed control only with no ability to reverse the motor. Another disadvantage is that I have seen reports of the controller overheating, even to the point of smoking. In addition, the controller is drawing power in ways that a simple relay or even transistor would not, because all current is flowing through semiconductors. Another disadvantage is that they are not quite as versatile, usually having lower maximum voltage for the motor than the transistors I have described in other chapters. However, for the benefit of people who do not want to bother with individual components, I will explain how to use motor controllers. They are particularly useful when using stepper motors, which I will discuss in future chapters, because these require four or more connections and can save you time and effort configuring four or more transistors.

Figure 5.1 shows the L298V Dual H Bridge.

Out 1
Out 2
Out 4
Out 3

V+ G 5V A 1 2 3 4 B

Figure 5.1

You can see that this has five screw type connectors, labeled Out 1, Out 2, Out 3, Out 4, V+, G, and 5V. These allow you to make firm connections to wires. To connect a wire, you turn the screw counterclockwise, which opens the contact space below the screw. You then insert the wire into the space, and then turn the screw clockwise until the space closes and firmly grasps the wire. To connect a DC motor, connect one of the two motor wires to Out 1 and the other to Out 2. If you have a second DC motor to connect, connect one to he leads to Out 3 and the other to Out 4.

For power, connect the positive lead of your power supply to the V+ screw terminal and the negative to the G (for Ground) screw terminal. The power can be up to 35 volts. However, if the power supply is over 12 volts, you

MUST remove the jumper connector just above V+ and G connectors, beside the Out 1 and Out 2 connectors. This is very important. If your external power supply is 12 volts or less, leave this jumper connector in. If the motor power supply connected to V+ and G is over 13 volts, remove it. This brings us to the 5V screw connector. If the jumper connector is connected, this is a 5-volt power source coming from the motor controller that you can use to power 5 volt circuits, even the Arduino itself by connecting it to the Arduino 5V pin. If you do this, you do not need an additional power supply to the Arduino. If you have removed the jumper, it is a 5-volt INPUT, which means you need to supply 5 volts to this connect to power the digital circuitry of the L298N. The easiest way to do that is to connect the 5V pin of your Arduino to this screw connector. However, if you do this, you need a separate power supply for your Arduino. Note that if you have a power supply to the L298N that is 12 volts or less (leaving the jumper in place) and you have a separate power supply to the Arduino, you do not need to connect the 5V connector on the L298N to anything.

    Now we come to the pins labeled A, 1, 2, 3, 4 and B on Figure 5.1. These are for connecting to the Arduino digital output pins. First, notice that there is a jumper connected to A and another jumper connected to B. If you are using DC motors, you must remove those jumpers. Do not discard them, however, because you will need them if you later want to use the L298N for stepper motors. Once you have removed the jumpers, you will see A and B pins similar to pins 1 through 4. These are your speed control pins. You should connect the front A pin of the two pins that were under the A jumper to an PWM digital output pin on the Arduino. This will control the speed of a motor connected to OUT 1 and OUT 2. You should connect the front B pin of the two pins that were under the B jumper to another PWM digital output pin on the Arduino. This will control the speed of a motor connected to OUT 3 and OUT 4. This assumes that you are going to use two motors. If you are using only on DC motor, you only need to connect one of the pins. It also assumes that you want to control the speed of the motors. If you only want to be able to turn the motors on and off, you can connect the A and B pins to any digital output. Just as you did with the transistors in chapters 3 and 4, you will control the speed of the motors by controlling the pulse width of the PWM pins, where using analogWrite(pin,0) stops the motor completely, analogWrite(pin,255) causes the motor to go full speed, and levels in between cause proportionally slower speeds.

Pins 1 and 2 control the direction of the motor connected to OUT 1 and OUT 2. One should be HIGH and one should be low. If pin 1 is HIGH and pin 2 is LOW, the motor will go one way. If pin 1 is LOW and pin 2 is HIGH, it will go the other way. Likewise, pins 3 and 4 control the direction of the motor connected to OUT 3 and OUT 4. If pin 3 is HIGH and pin 4 is LOW, the motor connected to OUT 3 and OUT 4 will go one way. If pin 3 is LOW and pin 4 is HIGH, that motor goes the other way. Thus, between the three pins for each motor, you can control speed and direction of each motor. If you only want to have the motor go in one direction, you could save two digital output pins for other uses by connecting either 1 or 2 to 5V and the other to GND, and doing the same with pins 3 and 4. You would just have to make sure you did not have both pins connected to either GND or 5V.

Sketch 5.1 can be used to control the speed and direction of two DC motors. As in previous chapters, I will use the serial interface to send commands to demonstrate the principle. Because of the two motors, the commands are a little more complicated. If you type A and press ENTER in the serial monitor, all following commands will go to motor A until you input B. Once you input B and press ENTER, all commands will go to motor B. Once you input F (for forward) and press ENTER, the motor will be in forward mode. Once you input R and press ENTER, the motor will be in reverse mode. Note that the motor will still not start until you input a speed. Once you input a number from 0 to 100, the current motor (A or B) will start moving at that speed. You can change speed or direction at any time by inputting another number or B or R. You can change the motor you are sending commands to at any time by inputting A or B at any time. OF course, as usual, this is just for demonstration purposes. You will probably want to have some more interesting way to control the motors. Here is Sketch 5.1.

```
#define AAnalogOutPin 9
#define AMotorPin1 8
#define AMotorPin2 7
#define BAnalogOutPin 11
#define BMotorPin1 12
#define BMotorPin2 13

String motorCommand;
String motorID;
int speedVar;
bool isNumber;

void setup() {
  // put your setup code here, to run once:
  pinMode(AAnalogOutPin, OUTPUT);
  analogWrite(AAnalogOutPin, 0);
  pinMode(BAnalogOutPin, OUTPUT);
  analogWrite(BAnalogOutPin, 0);
  pinMode(AMotorPin1, OUTPUT);
  pinMode(AMotorPin2, OUTPUT);
  setMotorAForward();
  pinMode(BMotorPin1, OUTPUT);
  pinMode(BMotorPin2, OUTPUT);
  setMotorBForward();
  Serial.begin(9600);
} // End of setup

void loop() {
  // put your main code here, to run repeatedly:
```

```
    if (Serial.available()>0) {
      motorCommand = Serial.readString();
      motorCommand.toUpperCase();
      isNumber = false;
      if (motorCommand >= "0" && motorCommand <=
"9") {isNumber = true;}
      if (isNumber) {
        speedVar = motorCommand.toInt();
        speedVar = map(speedVar,0,100,0,255);
      }
      Serial.println (motorCommand);
      if (motorCommand == "A" || motorCommand == "B")
{motorID = motorCommand;}

      if (motorCommand == "R" && motorID == "A") {
        setMotorABackward();
      }
      if (motorCommand == "F" && motorID == "A") {
        setMotorAForward();
      }
      if (motorCommand == "R" && motorID == "B") {
        setMotorBBackward();
      }
      if (motorCommand == "F" && motorID == "B") {
        setMotorBForward();
      }
      if (isNumber && motorID == "A") {
        setMotorASpeed();
      }
      if (isNumber && motorID == "B") {
        setMotorBSpeed();
      }
    } // End of if statement
} //End loop

void setMotorAForward(){
  digitalWrite(AMotorPin1, LOW);
  digitalWrite(AMotorPin2, HIGH);
```

```
}

void setMotorABackward(){
  digitalWrite(AMotorPin1, HIGH);
  digitalWrite(AMotorPin2, LOW);
}

void setMotorBForward(){
  digitalWrite(BMotorPin1, LOW);
  digitalWrite(BMotorPin2, HIGH);
}

void setMotorBBackward(){
  digitalWrite(BMotorPin1, HIGH);
  digitalWrite(BMotorPin2, LOW);
}

void setMotorASpeed(){
  analogWrite(AAnalogOutPin, speedVar);
}

void setMotorBSpeed(){
  analogWrite(BAnalogOutPin, speedVar);
}
```
Sketch 5.1

I start out by defining the analog and digital pins for motor A and B. A new variable defines is motorID, which will hold the ID of the motor currently being controlled (A or B).

In order to make it easier for you to adapt this to your own uses, I have put some of the code in subroutines. The subroutine setMotorAForward configures motor A to go forward. You can see that it sets AMotorPin1 to LOW and AMotorPin2 to HIGH. The subroutine setMotorABackward configures motor A to go in reverse by setting AMotorPin1 to HIGH and AMotorPin2 to LOW. The subroutine setMotorBForward configures motor A to go forward. You can see that it sets BMotorPin1 to LOW and BMotorPin2 to HIGH. The subroutine setMotorBBackward configures motor A to go in reverse by setting BMotorPin1 to HIGH and BMotorPin2 to LOW. I should note at this point that forward and backward are arbitrary in this discussion. What actually constitutes forward and backward in your application will depend on how you mount your motors. You can simply change the names of the subroutines if you are getting the reverse of the effect you want. You can also simply reverse the motor lead to OUT 1 and 2 or the leads to OUT 3 and 4.

The subroutine setMotorASpeed sets the speed of motor A. The subroutine setMotorBSpeed sets the speed of motor B.

The setup routine sets all the digital pins used to OUTPUT mode. It also calls the subroutines to set those pin to forward mode and sets the speed to 0, to ensure the motors are ready but not actually running when the Arduino starts up.

The loop routine is similar to the ones in previous sketches. It input the command from the serial monitor and capitalizes it. It then checks to see if the command is a number. It sets the variable isNumber to false, then sets it to true if the first character is between 0 and 9. The if section after that acts on the command. If the command is a number, it is stored in speedVar. If it is A or B, the string variable motorID is set to that value. If it is R or F (Reverse or Forward), the subroutine for the current motor, A or B, is called to set that motor to reverse or forward. If the command was a number, the current motor is set to that speed. The main loop of the sketch is not really very important, since it is just for demonstration purposes. The important part of this sketch is the subroutines, which you can incorporate into your own sketch and call based on whatever method you are using to command the motors, such as a remote-control device.

As noted earlier in this chapter, you may not need all these pins and subroutines. If you have no desire to control the speed, you can simply connect the A and B pins on the shield to 5 volts, and leave defining the speed pins, the initializing of the speed pins, and the speed subroutines, out of your sketch. If you only want the motors to go in one direction, you can skip defining the motor pins, initializing them, and the direction subroutines out of your sketch.

Chapter 6

Unipolar Stepper Motors

As explained in previous chapters, DC motors go as fast as they can for the voltage you supply. You can control their speed by controlling the voltage applied, or by stopping and starting the voltage as described in previous chapters. However, this control is not very precise, since it also depends on the amount of resistance the motor is encountering at any given moment.

Unipolar stepper motors work on a different principle. They usually have five control wires. You apply a series of pulses to these wires, and each time you send a complete series of pulses, the motor rotates one step. Each step can be very small, often as little as 1.8 degrees, although in some motors a step can be as much as a quarter of a rotation. The most common step sizes are 1.8, 7.5, and 15 degrees. However, many stepper motors also have step down gear boxes that result in even smaller steps. Such stepper motors often have such small steps that you cannot even see one step with the naked eye.

The main advantage of stepper motors is that you can have extremely precise control over how far and how fast the shaft rotates. They are very good for devices where you need to very precisely move something and know exactly how far you have moved it, such as a 3-D printer, CNC machine, laser cutter, or other precision tools. Another popular use is robot arms or fingers, where you want to be able to precisely control the position of the arm or how far you close the hand. The main disadvantage of stepper motors is that they tend to rotate slower. They are also much more complicated to wire and to program code for. You must apply power in complicated patterns to four of the five wires to cause the rotation. Thus, stepper motors are used in specific cases where control and precision are important.

There are two types of stepper motors: unipolar and bipolar. The name refers to the fact that in unipolar motors, the current always flows in the same direction in the coils, while in bipolar motors the current reverses direction. The main advantage of bipolar motors is that they can be more powerful, creating more torque. However, they require more complicated circuitry to operate because you have to reverse the polarity of the coils. In this chapter, I will discuss unipolar stepper motors.

Unipolar stepper motors have four coils. They usually have five wires connected to these coils. Figure 6.1 shows the internal wiring of a unipolar stepper motor.

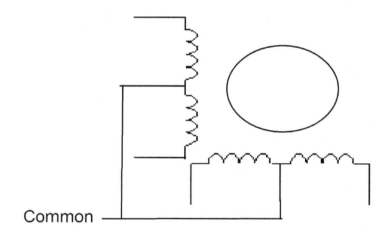

Figure 6.1

The common wire is connected to one side of the power source, almost always the positive. Power is applied to this wire at all times when the motor is running. The other four wires are connected to the negative end of the power supply. Power is turned on and off for each of these wires in several different combinations. Each combination energizes different configurations of coils to move the motor one step forward or backward. The combinations must be performed in a set order. The combinations then repeat in a cycle when it reaches the last combination.

In a full step cycle (I will discuss half step cycles later), there are four combinations to each cycle. Here are the combinations for one cycle to cause the motor to turn one way.

Step 1	Off	Off	On	On
Step 2	Off	On	On	Off
Step 3	On	On	Off	Off
Step 4	On	Off	Off	On

Table 6.1

You can cause the motor to turn the opposite direction by using the following combination.

Step 1	On	On	Off	Off
Step 2	Off	On	On	Off
Step 3	Off	Off	On	On
Step 4	On	Off	Off	On

Table 6.2

Power can be turned off by electronically "disconnecting" the wire, such as shutting off current flow with a transistor. Current flow can also be cut off by setting the voltage to that wire to the same positive voltage level as the main wire, since current will not flow between two points at the same voltage level.

Figure 6.2. shows a circuit to control the motor, using transistors, resistors, and diodes. As discussed in Chapter 3, the diodes are to handle the inductive flyback of the motor. Since the motor has four coils, you need four diodes.

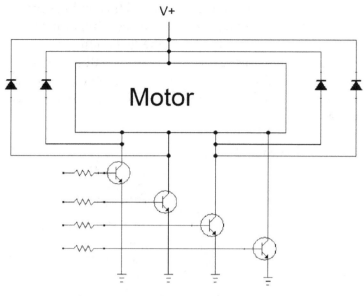

Figure 6.2

    This shows the main connector to the V+ power and the four control wires being controlled by the transistors. The resistors would be connected to four Arduino digital outputs.

    However, this is a lot of wiring to do, and most people take the easier path of using a motor controller board. There are many controller boards available, with many advantages and disadvantages such as price, the amount of power they can handle, how many motors they can control, etc. I will discuss several popular models in this chapter.

    One motor controller popular with Arduino hobbyists is the ULN2003 shown in Figure 6.3.

Figure 6.3

On the left, you can see male pins labelled IN 1 through IN 4. You connect each of these to one Arduino digital output.

At the bottom you can see − and + male connectors. These are the power supply inputs for the motor. The − must be connected to ground on your Arduino. If you are using a low current 5-volt motor, you can connect the motor controller pin labelled + to your Arduino 5-volt output. If you are using a higher voltage motor, it gets a little trickier. Suppose, for example, you want to run a 12-volt stepper motor. You will, of course, need a 12-volt power supply. You have several options. You can connect the power supply negative to both the − pin on the motor controller and the Arduino GND and the power supply 12 volt + to the motor controller and also the Arduino VIN to power the Arduino. That would power both the motor controller and the Arduino from the 12-volt power supply. Similarly, you can power the Arduino from the 12-volt power supply through the power jack on the Arduino and then connect the VIN Arduino pin to the motor controller + pin. However, this will supply a little less than 12 volts to the motor controller, since there is about a .7 volt drop across the diode inside the Arduino from the jack to the VIN. You can also power the Arduino and motor controller completely separately, connecting the 12-volt power supply positive to the motor controller – and + and the Arduino GND to the motor controller –. In this case, it is not necessary to connect the Arduino 5V or VIN to the motor controller. Power just flows from the IN1 through IN 4 to the -.

In Figure 6.3 you can also see five pins on the right. These are where you connect the motor. Figuring out which wire to connect to which pin could be a challenge. Fortunately, most motors you would run from this controller come with a plug that fits directly onto the five pins. Figure 6.4 shows the motor.

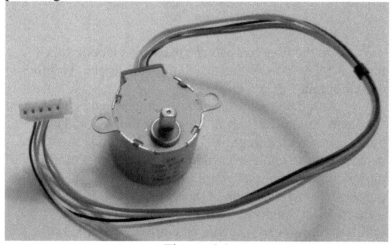

Figure 6.4

You can see that the five holes on the motor cable will fit onto the five pins on the motor controller. The cable connector even comes notched so that you cannot insert it in the wrong direction. These motors come in 5 and 12 volt versions.

With this controller, the fifth pin (the one in front of the picture) supplies the positive voltage. Each of the other pins goes negative when a 5 volt positive signal is supplied to the corresponding IN pin. For example, if you apply a positive 5 volts to IN1, the voltage on motor pin A goes negative. If you apply a ground current to IN1, motor pin A goes positive. Remember that the common pin is positive, so causing the motor pin to go positive actually stops current from flowing, while causing it to go negative (ground) allows current to flow. Sketch 6.1 shows how to use this with an Arduino. It allows you to input a positive or negative number from the serial monitor. The motor will rotate forward that number of steps if the number is positive and backward that number of steps if it is negative.

```
#define MotorWireOff LOW
#define MotorWireOn HIGH

int MotorPin1 = 8;
int MotorPin2 = 9;
int MotorPin3 = 10;
int MotorPin4 = 11;
int MotorDelay = 3;
int motorStep = 1;
int control;

void setup() {
//declare the motor pins as outputs
 pinMode(MotorPin1, OUTPUT);
 pinMode(MotorPin2, OUTPUT);
 pinMode(MotorPin3, OUTPUT);
 pinMode(MotorPin4, OUTPUT);
 ShutDown();
 Serial.begin(9600);
}

void loop(){
 if (Serial.available() > 0) {
```

```
    control = Serial.parseInt();
    Serial.print("Control = ");
    Serial.println(control);
    if (control > 0) {
      for (int i=1; i <= control; i++){
        MotorForward();
      } // End for
    }

    if (control < 0) {
      control = abs(control);
      for (int i=1; i <= control; i++){
        MotorBackward();
      } // End for
    }
    ShutDown();
  }
}

void MotorForward(){
 if (motorStep == 1) {
  digitalWrite(MotorPin1, MotorWireOff);
  digitalWrite(MotorPin2, MotorWireOff);
  digitalWrite(MotorPin3, MotorWireOn);
  digitalWrite(MotorPin4, MotorWireOn);
 }
 if (motorStep == 2) {
   digitalWrite(MotorPin1, MotorWireOff);
   digitalWrite(MotorPin2, MotorWireOn);
   digitalWrite(MotorPin3, MotorWireOn);
   digitalWrite(MotorPin4, MotorWireOff);
 }
 if (motorStep == 3) {
   digitalWrite(MotorPin1, MotorWireOn);
   digitalWrite(MotorPin2, MotorWireOn);
   digitalWrite(MotorPin3, MotorWireOff);
   digitalWrite(MotorPin4, MotorWireOff);
 }
```

```
  if (motorStep == 4) {
    digitalWrite(MotorPin1, MotorWireOn);
    digitalWrite(MotorPin2, MotorWireOff);
    digitalWrite(MotorPin3, MotorWireOff);
    digitalWrite(MotorPin4, MotorWireOn);
  }
  delay (MotorDelay);
  ++motorStep;
  if (motorStep > 4) {motorStep = 1;}
}

void MotorBackward (){
  if (motorStep == 4) {
    digitalWrite(MotorPin1, MotorWireOn);
    digitalWrite(MotorPin2, MotorWireOn);
    digitalWrite(MotorPin3, MotorWireOff);
    digitalWrite(MotorPin4, MotorWireOff);
  }
  if (motorStep == 3) {
    digitalWrite(MotorPin1, MotorWireOff);
    digitalWrite(MotorPin2, MotorWireOn);
    digitalWrite(MotorPin3, MotorWireOn);
    digitalWrite(MotorPin4, MotorWireOff);
  }
  if (motorStep == 2) {
    digitalWrite(MotorPin1, MotorWireOff);
    digitalWrite(MotorPin2, MotorWireOff);
    digitalWrite(MotorPin3, MotorWireOn);
    digitalWrite(MotorPin4, MotorWireOn);
  }
  if (motorStep == 1) {
    digitalWrite(MotorPin1, MotorWireOn);
    digitalWrite(MotorPin2, MotorWireOff);
    digitalWrite(MotorPin3, MotorWireOff);
    digitalWrite(MotorPin4, MotorWireOn);
  }
  delay (MotorDelay);
  --motorStep;
```

```
  if (motorStep < 1) {motorStep = 4;}
}

void ShutDown(){
 digitalWrite(MotorPin1, MotorWireOff);
 digitalWrite(MotorPin2, MotorWireOff);
 digitalWrite(MotorPin3, MotorWireOff);
 digitalWrite(MotorPin4, MotorWireOff);
}
```
Sketch 6.1

    In order to make it easier for you to take out the code that you want to use for your own projects, I have put the actual motor control code into subroutines that you can easily copy and paste into your own sketches. The MotorForward subroutine causes the motor to go in one direction, which we will call forward. The MotorBackward subroutine turns the motor in the opposite direction. (Of course, what is "backward" and "forward" depends on the motor orientation in your project, but the terms are convenient.) If you look at Tables 6.1 and 6.2 above, you will see that MotorForward turns the motor coils on and off in the sequence given in Table 6.1, and MotorBackward does that as given in Table 6.2. The constant MotorWireOff is defined at the beginning of the sketch as the Arduino output value that will turn off power to that motor wire. The constant MotorWireOn is defined at the beginning of the sketch as the output value that will turn on power to that motor wire. Since for this motor controller, setting the Arduino output to an IN pin on the controller HIGH sets the motor pin LOW, allowing current to flow through the coil, MotorWireOn is defined as HIGH. Likewise, sending the Arduino output to an IN pin on the controller LOW sends the motor controller output positive, shutting off power through the coils, so MotorWireOff is defined as LOW.

    Each subroutine applies power to the four coils in sequence. In order to allow for the great precision, I have set them up so that each call to one of these subroutines advances one step. Since it is necessary to perform the steps in the proper order, the variable motorStep keeps track of which step to perform. If motorStep is 1, step 1 is performed. If motorStep is 2, step 2 is performed, and so on. In the MotorForward subroutine, after the step is performed, motorStep is increased by 1 using the ++ motorStep command. (The code ++ followed by a numeric variable causes that variable to increase by 1.) In the MotorBackward subroutine, after the step is performed, motorStep is decreased by 1 using the -- motorStep command, because the motor is moving in the opposite direction so the current motor position is decreasing. The motorStep variable is initially set to 1 when it is defined at the beginning of the sketch.

Note that after each step, there is a delay. It is absolutely necessary to have a delay to allow the motor to actually respond to the current power configuration. Applying power to some of the coils pulls the motor shaft in that direction. It takes time for the motor shaft to actually rotate to that position. If you do not delay between changes in the power configuration, the motor will not move. The best delay time can vary with different motors, so I have used an integer variable, MotorDelay, that you can set at the beginning of your sketch. I find that about 3 milliseconds works well for many motors, so MotorDelay would be set to 3. A delay of 1 millisecond will not be enough, and the motor will not move at all. I found that a delay of 2 allows the motor to turn most of the time, but occasionally the motor will jam with a 2 millisecond delay.. Any large delay will work, but the motor will simply turn slower. Of course, you may need to experiment with your motor to find the best delay

You may have noted that after each step, some pins are HIGH and some LOW, so power is still flowing at the end of each step. This is fine if the subroutine is about to be called again immediately, but when you have turned the motor as much as you want to and you are not going to call the subroutine again, you should turn off all power. Note that the motor will not continue to turn even if you leave power on, because the motor actually advances when the power changes. However, leaving the power to any coils on will waste energy, drain your battery, and cause the coils to heat up. Therefore, you should turn off the power when you do not want the motor to be turning. You can do this with the ShutDown subroutine. This sets all Arduino digital outputs MotorWireOff, making the motor pins positive, cutting off power to each coil.

Each of the motor subroutines will advance the motor by a tiny amount, usually too little to see. In order to advance the motor by a significant amount, you need to call the subroutine repeatedly, generally with a for loop.

At the beginning of the sketch, I define MotorWireOn as HIGH and MotorWireOff as LOW, as explained previously. I also define the four digital output pins that will be used. I have arbitrarily used 8 through 11, but you can use any four pins you want. Of course, MotorPin1 is the pin that you will connect to IN1 on the motor controller, and so on. I define MotorDelay and set it to 3 as explained previously, and set motorStep to 1. The variable control will hold the number you sent from the serial monitor for the number of steps to take.

The setup routine defines the pins as output and initializes the serial port for communications. It also calls the ShutDown subroutine to make sure that no power is going to the motor when the program starts. This is just so that the motor is not drawing (and wasting) power until you are ready to use it.

In the loop routine, if a message is available at the serial port, the line control = Serial.parseInt(); inputs that message and converts it to an integer and stores that value in control. The Serial.print statements simply echo it back to confirm that the message was received correctly. Using if commands to determine if control is greater than or less than 0, if the value of control is greater than 0, the four loop calls MotorForward the number of times assigned to control. If control is less than 0, the abs function converts control to its positive value and then a for loop calls the MotorBackward subroutine that number of times. In either case, the ShutDown subroutine is then called to power down the coils.

As I mentioned, the main advantage of stepper motors is to allow you to make very carefully controlled small steps. The tables and motor subroutines discussed above have dealt with taking single steps. However, if you really want to make extremely small steps, you can do half steps by turning off power to one coil at a time instead of two. Sketch 6.2 is a small variation of Sketch 6.1 that does this.

```
#define MotorWireOff LOW
#define MotorWireOn HIGH

int MotorPin1 = 8;
int MotorPin2 = 9;
int MotorPin3 = 10;
int MotorPin4 = 11;
int MotorDelay = 2;
int motorStep = 1;
int control;

void setup() {
//declare the motor pins as outputs
  pinMode(MotorPin1, OUTPUT);
  pinMode(MotorPin2, OUTPUT);
  pinMode(MotorPin3, OUTPUT);
  pinMode(MotorPin4, OUTPUT);
  ShutDown();
  Serial.begin(9600);
}

void loop(){
if (Serial.available() > 0) {
  control = Serial.parseInt();
```

```
  MotorDelay = 1;
  Serial.print("Control = ");
  Serial.println(control);
  if (control > 0) {
    for (int i=1; i <= control; i++){
      MotorForward();
      //Serial.print("step ");
      //Serial.println(i);
    } // End for
  }

  if (control < 0) {
    control = abs(control);
    for (int i=1; i <= control; i++){
      MotorBackward();
      //Serial.print("step ");
      //Serial.println(i);
    } // End for
  }
  ShutDown();
  }
}

void MotorForward(){
if (motorStep == 1) {
digitalWrite(MotorPin4, MotorWireOn);
digitalWrite(MotorPin3, MotorWireOff);
digitalWrite(MotorPin2, MotorWireOff);
digitalWrite(MotorPin1, MotorWireOff);
}
if (motorStep == 2) {
digitalWrite(MotorPin4, MotorWireOn);
digitalWrite(MotorPin3, MotorWireOn);
digitalWrite(MotorPin2, MotorWireOff);
digitalWrite(MotorPin1, MotorWireOff);
}
if (motorStep == 3) {
digitalWrite(MotorPin4, MotorWireOff);
```

```
digitalWrite(MotorPin3, MotorWireOn);
digitalWrite(MotorPin2, MotorWireOff);
digitalWrite(MotorPin1, MotorWireOff);
}
if (motorStep == 4) {
digitalWrite(MotorPin4, MotorWireOff);
digitalWrite(MotorPin3, MotorWireOn);
digitalWrite(MotorPin2, MotorWireOn);
digitalWrite(MotorPin1, MotorWireOff);
}
if (motorStep == 5) {
digitalWrite(MotorPin4, MotorWireOff);
digitalWrite(MotorPin3, MotorWireOff);
digitalWrite(MotorPin2, MotorWireOn);
digitalWrite(MotorPin1, MotorWireOff);
}
if (motorStep == 6) {
digitalWrite(MotorPin4, MotorWireOff);
digitalWrite(MotorPin3, MotorWireOff);
digitalWrite(MotorPin2, MotorWireOn);
digitalWrite(MotorPin1, MotorWireOn);
}
if (motorStep == 7) {
digitalWrite(MotorPin4, MotorWireOff);
digitalWrite(MotorPin3, MotorWireOff);
digitalWrite(MotorPin2, MotorWireOff);
digitalWrite(MotorPin1, MotorWireOn);
}
if (motorStep == 8) {
digitalWrite(MotorPin4, MotorWireOn);
digitalWrite(MotorPin3, MotorWireOff);
digitalWrite(MotorPin2, MotorWireOff);
digitalWrite(MotorPin1, MotorWireOn);
}
delay(MotorDelay);
 ++motorStep;
 if (motorStep > 8) {motorStep = 1;}
}
```

```
void MotorBackward (){
if (motorStep == 8) {
digitalWrite(MotorPin1, MotorWireOn);
digitalWrite(MotorPin2, MotorWireOff);
digitalWrite(MotorPin3, MotorWireOff);
digitalWrite(MotorPin4, MotorWireOff);
}
if (motorStep == 7) {
digitalWrite(MotorPin1, MotorWireOn);
digitalWrite(MotorPin2, MotorWireOn);
digitalWrite(MotorPin3, MotorWireOff);
digitalWrite(MotorPin4, MotorWireOff);
}
if (motorStep == 6) {
digitalWrite(MotorPin1, MotorWireOff);
digitalWrite(MotorPin2, MotorWireOn);
digitalWrite(MotorPin3, MotorWireOff);
digitalWrite(MotorPin4, MotorWireOff);
}
if (motorStep == 5) {
digitalWrite(MotorPin1, MotorWireOff);
digitalWrite(MotorPin2, MotorWireOn);
digitalWrite(MotorPin3, MotorWireOn);
digitalWrite(MotorPin4, MotorWireOff);
}
if (motorStep == 4) {
digitalWrite(MotorPin1, MotorWireOff);
digitalWrite(MotorPin2, MotorWireOff);
digitalWrite(MotorPin3, MotorWireOn);
digitalWrite(MotorPin4, MotorWireOff);
}
if (motorStep == 3) {
digitalWrite(MotorPin1, MotorWireOff);
digitalWrite(MotorPin2, MotorWireOff);
digitalWrite(MotorPin3, MotorWireOn);
digitalWrite(MotorPin4, MotorWireOn);
}
```

```
if (motorStep == 2) {
digitalWrite(MotorPin1, MotorWireOff);
digitalWrite(MotorPin2, MotorWireOff);
digitalWrite(MotorPin3, MotorWireOff);
digitalWrite(MotorPin4, MotorWireOn);
}
if (motorStep == 1) {
digitalWrite(MotorPin1, MotorWireOn);
digitalWrite(MotorPin2, MotorWireOff);
digitalWrite(MotorPin3, MotorWireOff);
digitalWrite(MotorPin4, MotorWireOn);
}
 delay (MotorDelay);
 --motorStep;
 if (motorStep < 1) {motorStep = 8;}
}

void ShutDown(){
 digitalWrite(MotorPin4, MotorWireOff);
 digitalWrite(MotorPin3, MotorWireOff);
 digitalWrite(MotorPin2, MotorWireOff);
 digitalWrite(MotorPin1, MotorWireOff);
}
```

Sketch 6.2

Comparing this to Sketch 6.1, you can see that it is basically the same except that when it changes the connections to the motor coils, it does it one coil at a time instead of changing two coils at a time. This means that instead of always having two coils energized, half the time it only has one coil energized. The advantage is that you get twice the resolution in positioning the motor. The disadvantage is that you get an average torque about 2/3 as much. Thus, the half step sketch is good if you really need to make very small movements and the object you are moving offers very little resistance. Note that this is entirely a difference in the programming code. It is the same motor. You might also note that you can often use a shorter delay time (MotorDelay) operating in half step mode.

Another motor controller you can use is the one discussed in chapter 5, the L298N. For your convenience, Figure 6.5 shows this again.

Figure 6.5

To use this for stepper motors, you must leave the jumpers A and B in place as shown in the photo. If you are using a low current (under .5 amps) 5-volt motor, you can connect V+ to your Arduino 5V and G to your Arduino GND to power the motor controller. If you are using a higher current or voltage motor, connect the positive terminal of an external power supply (up to 35 volts) to V+ and the negative to G. Also connect the G to your Arduino GND. If you are using a power supply 12 volts or under, you can use the 5V connection to power a 5-volt device. For example, you can connect 5V to your Arduino 5V. If you are using a power supply over 12 volts, disconnect the jumper next to the OUT2 connection and DO NOT connect anything to the 5V.

Connect your Arduino digital output pins 8 through 11 (assuming these are the digital output pins you want to use) to the pins labeled 1, 2, 3, and 4 on the diagram. Connect the common wire of the motor to V+. Connect the other four wires of the motor to OUT1 through OUT 4. The wires generally come out of the motor in a row, with the common wire in the middle. Connect the two wires on one side of the center wire to OUT1 and OUT2, with the one closest to the center going to OUT1. Connect the two wires on the other side of the center wire to OUT3 and OUT4, with the wire closest to the center going to OUT3. If the motor does not run with this configuration, swap the wires going to OUT1 and OUT2.

Now for the software. The sketches 6.1 and 6.2 will work, but with one very important exception. With the ULN2003 driver shown in Figure 6.3, the output to the motor goes high (positive) when the input at the in pins goes low. With the L298N, the output to the motor goes positive when the input goes positive. This is most important in the ShutDown subroutine, because it actually means that power would be applied to all four coils all the time after you run that subroutine. The motor will not turn because the power is not switching, but the motor will get very hot and you will drain your batteries (or simply waste power). Because of this, you want to reverse the HIGH's and LOWs in the sketch. This is why I defined MotorWireOn and MotorWireOff and used these in Sketches 6.1 and 6.2, rather than simply putting HIGH and LOW in the digitalWrite commands. It makes it much easier to alter the sketches for different motor controllers. In order to change these sketches to work with a motor controller like the L298N, simply change the first two define lines to read
#define MotorWireOff HIGH
#define MotorWireOn LOW

Sketches 6.1 and 6.2 allowed you to tell the motor to go a certain number of steps or half steps. Suppose you want to have the motor continue running at a certain speed in one direction or the other until you tell it to stop. Sketch 6.3 shows how to do that. It allows you to send a number from -100 to 100 from the serial monitor. Sending -100 causes it to go maximum speed in reverse, 100 causes it to go maximum speed forward, 0 causes the motor to stop, and numbers like 99 cause the motor to go at partial speed. The speed decreases very quickly when you reduce the speed from 100. For example, 97 is about half as fast as 100, and 91 is about 25% of 100. Numbers outside the range of -100 to 100 can cause erratic results, such as causing the program to freeze for long periods of time.

Because you can use either single step or half step subroutines in this sketch, I have left out the actual MotorForward, MotorBackward, and ShutDown subroutines from Sketch 6.3. I am including only the initialization, setup, and loop routines so you can then paste in the actual motor control subroutines you want to use. This sketch is about how you control the subroutines.

```
//Use next two if motor controller output is opposite from input
#define MotorWireOff LOW
#define MotorWireOn HIGH
//Use next two if motor controller output is same as input
//#define MotorWireOff HIGH
//#define MotorWireOn LOW


int MotorPin1 = 8;
int MotorPin2 = 9;
int MotorPin3 = 10;
int MotorPin4 = 11;
int MotorDelay = 3;
int motorStep = 1;
int control;
int MaxSpeed = 100;
```

```
void setup() {
//declare the motor pins as outputs
pinMode(MotorPin1, OUTPUT);
pinMode(MotorPin2, OUTPUT);
pinMode(MotorPin3, OUTPUT);
pinMode(MotorPin4, OUTPUT);
ShutDown();
Serial.begin(9600);
}

void loop(){
if (Serial.available() > 0) {
  control = Serial.parseInt();
  Serial.print("Control = ");
  Serial.println(control);
  }
  if (control > 0) {
     MotorForward();
     delay(MaxSpeed - control);
    }
  if (control == 0) {ShutDown();}
  if (control < 0) {
     MotorBackward();
     delay(MaxSpeed - abs(control));
    }
}
```

Sketch 6.3

First, to make it easy to adapt this sketch, I have included define statements for MotorWireOff and MotorWireOn for both types of controller, and remarked out the #define MotorWireOff HIGH and #define MotorWireOn LOW statements. This allows you to use this sketch for either type of controller just by remarking out the appropriate statements. The only other change I made to the initialization is to add the integer MaxSpeed and set it to 100, so that 100 is the maximum speed you can set for the motor. The setup routine is the same as in Sketches 6.1 and 6.2.

In the loop, it first inputs the value of control, which should be between -100 and 100. One big difference from the previous sketches is that the part of the code that uses the value of control is not within the if (Serial.available() > 0) statement. In previous sketches, the for loop that acted on the value of control was only reached immediately after a new value of control was sent, and therefore was only acted on once. In this sketch, if no new value of control is sent, the loop continues to act on the current value until a new value is sent.

If control is greater than 0, the MotorForward is called and then the delay statement causes the program to pause by a time of MaxSpeed − control milliseconds. For example, if control is 97, the delay is 100 − 97 = 3 milliseconds. This might seem like a short time, but remember that it occurs between every step or half step the motor takes. There can be hundreds of steps in one rotation, and if the motor is geared down, it can be thousands.

If control is less than 0, the MotorBackward subroutine is called and then there is a delay of MaxSpeed minus the absolute (positive) value of control. Therefore, negative 100 is the fastest speed, with negative 97 being about half that speed. If control is exactly 0, the ShutDown subroutine is called.

After executing the call to the appropriate subroutine, the loop repeats. If no new command has been sent, this process occurs over and over using the same value of control. There is no need to send additional commands for additional movement, the way there was in Sketches 6.1 and 6.2. Thus, the stepper motor can be used either to generate either a carefully controlled amount of movement, or a carefully controlled speed of rotation.

Chapter 7

Bipolar Stepper Motors

Bipolar motors are similar to unipolar motors, in that you cause them to make small individual steps by applying power in complicated patterns to individual wires. However, the basic bipolar motor usually has four wires instead of five. The schematic is shown in Figure 7.1. Figure 7.2 shows two typical bipolar motors.

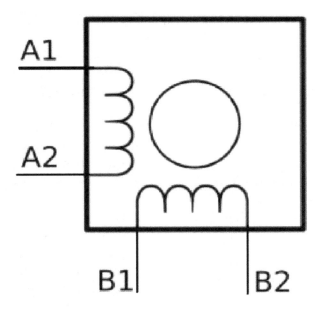

Figure 7.1

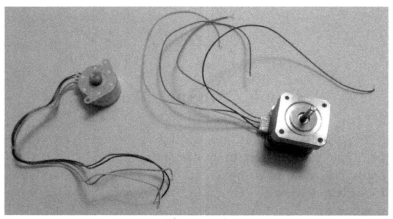

Figure 7.2

The name bipolar refers to the fact that, unlike unipolar motors, you do not just turn the current to the wires on and off. You sometimes reverse it. This can give the motor more torque (strength). However, it does require that you use that you use what is called an H bridge motor controller.

The L298N discussed in previous chapters is such a controller. Connecting the bipolar motor to this controller is almost the same as connecting the unipolar motor. The key is finding the right wires to connect. Referring to Figure 7.1, you connect A1 and A2 to Out1 and Out2, and B1 and B2 to Out3 and Out4. Note: You can also connect B1 and B2 to Out1 and Out2 and A1 and A2 to Out3 and Out4. It does not even matter if you reverse the order and connect A1 to Out2 and A2 to Out1, or B1 to Out4 and B2 to Out3. Reversing the wires generally has the effect of reversing the direction the motor spins for a given sequence of outputs from the Arduino. The important part is to make sure that you have the two wires to one coil going to Out1 and Out2, and the two wires from the other coil going to Out3 and Out4, but that is not really so hard. First, A1 and A2 are usually beside each other where they exit the motor, as are B1 and B2. If you are unsure, there is a very simple test. Just connect an ohmmeter to two of the wires. If they have finite resistance, usually about 5 to 15 ohms, they two wires go to the same coil. If they are not to the same coil, the resistance will be infinite.

As for the other motor controller connections, it is the same as in Chapter 6. Connect each of four outputs from the Arduino (such as 8 through 11) to the four inputs on the motor controller, 1 through 4 as shown on Figure 6.5. Connect the Arduino ground (GND) to G and the power supply to V+ and G.

The software is also almost the same as in Chapter 6. The difference is just the order in which you set the outputs HIGH and LOW. There are three basic ways you can sequence the output. These are called one phase, two phase, and half step. In one phase, only one coil is powered at a time. This uses less power but produces less torque. It can be useful if the motor is used in an application that does not require much force and you need to minimize power draw, such as a battery driven device where you want to conserve the battery. Two phase powers both coils at once and gives more torque. Half step gives you smaller steps, which gives you more precise control of the movement. In half step, sometimes one coil is powered and sometimes both are powered. One problem with it is that the amount of torque varies with different portions of the step, producing uneven torque. However, if you want very small steps and resistance on the motor is so small you do not need much torque, half step can be useful.

Sketch 7.1 demonstrates the one phase. As in many of the other sketches, it accepts a command from the serial monitor to advance a certain number of steps, which can be positive or negative.

```
//Use next two if motor controller output is opposite from input
//#define MotorWireHigh LOW
//#define MotorWireLow HIGH
//Use next two if motor controller output is same as input
#define MotorWireHigh HIGH
```

```
#define MotorWireLow LOW

int MotorPin1 = 8;
int MotorPin2 = 9;
int MotorPin3 = 10;
int MotorPin4 = 11;
int MotorDelay = 3;
int motorStep = 1;
int control;

void setup() {
//declare the motor pins as outputs
pinMode(MotorPin1, OUTPUT);
pinMode(MotorPin2, OUTPUT);
pinMode(MotorPin3, OUTPUT);
pinMode(MotorPin4, OUTPUT);
ShutDown();
Serial.begin(9600);
}

void loop(){
if (Serial.available() > 0) {
  control = Serial.parseInt();
  Serial.print("Control = ");
  Serial.println(control);
  if (control > 0) {
    for (int i=1; i <= control; i++){
      MotorForward();
     } // End for
   }

  if (control < 0) {
    control = abs(control);
    for (int i=1; i <= control; i++){
      MotorBackward();
     } // End for
    }
  ShutDown();
```

}
}

void MotorForward(){
 if (motorStep == 1) {
   digitalWrite(MotorPin1, MotorWireLow);
   digitalWrite(MotorPin2, MotorWireLow);
   digitalWrite(MotorPin3, MotorWireLow);
   digitalWrite(MotorPin4, MotorWireHigh);
 }
 if (motorStep == 2) {
   digitalWrite(MotorPin1, MotorWireLow);
   digitalWrite(MotorPin2, MotorWireHigh);
   digitalWrite(MotorPin3, MotorWireLow);
   digitalWrite(MotorPin4, MotorWireLow);
 }
 if (motorStep == 3) {
   digitalWrite(MotorPin1, MotorWireLow);
   digitalWrite(MotorPin2, MotorWireLow);
   digitalWrite(MotorPin3, MotorWireHigh);
   digitalWrite(MotorPin4, MotorWireLow);
 }
 if (motorStep == 4) {
   digitalWrite(MotorPin1, MotorWireHigh);
   digitalWrite(MotorPin2, MotorWireLow);
   digitalWrite(MotorPin3, MotorWireLow);
   digitalWrite(MotorPin4, MotorWireLow);
 }
 delay (MotorDelay);
 ++motorStep;
 if (motorStep > 4) {motorStep = 1;}
}

void MotorBackward (){
 if (motorStep == 4) {
   digitalWrite(MotorPin1, MotorWireHigh);
   digitalWrite(MotorPin2, MotorWireLow);
   digitalWrite(MotorPin3, MotorWireLow);
```
```

```
  digitalWrite(MotorPin4, MotorWireLow);
}
if (motorStep == 3) {
  digitalWrite(MotorPin1, MotorWireLow);
  digitalWrite(MotorPin2, MotorWireLow);
  digitalWrite(MotorPin3, MotorWireHigh);
  digitalWrite(MotorPin4, MotorWireLow);
}
if (motorStep == 2) {
  digitalWrite(MotorPin1, MotorWireLow);
  digitalWrite(MotorPin2, MotorWireHigh);
  digitalWrite(MotorPin3, MotorWireLow);
  digitalWrite(MotorPin4, MotorWireLow);
}
if (motorStep == 1) {
  digitalWrite(MotorPin1, MotorWireLow);
  digitalWrite(MotorPin2, MotorWireLow);
  digitalWrite(MotorPin3, MotorWireLow);
  digitalWrite(MotorPin4, MotorWireHigh);
}
delay (MotorDelay);
--motorStep;
if (motorStep < 1) {motorStep = 4;}
}

void ShutDown(){
 digitalWrite(MotorPin1, MotorWireLow);
 digitalWrite(MotorPin2, MotorWireLow);
 digitalWrite(MotorPin3, MotorWireLow);
 digitalWrite(MotorPin4, MotorWireLow);
}
```

Sketch 7.1

This starts out by defining MotorWireHigh and MotorWireLow as HIGH and LOW. This is basically the same as the previous defining of MotorWireOn and MotorWireOff, with a name change to emphasize that for a bipolar motor, the voltage actually goes high and low to reverse the current, not just shut it off. Otherwise, initialization, setup, and the main loop are all the same as Sketches 6.1 and 6.2. The differences in this sketch are the MotorForward, MotorBackward, and ShutDown subroutines. For bipolar stepper motors, a coil is energized is the two wires are different (one high and one low), rather than if the wire is low. You might notice that in this sketch, in MotorForward and MotorBackward only one wire is high at a time. This is because, as mentioned before, in the one phase setup, only one coil is energized at a time. In ShutDown, all wires are driven low. However, since coils are energized by having the two wires to a coil be different, I could just as effectively have had all wires go high. In either case, no current would flow through the coils.

  To use the motor as a two phase system, change the MotorForward and MotorBackward subroutines in Sketch 7.1 to those in Sketch 7.2. To save space, making this book smaller (and cheaper), I will list only the subroutines that you need to replace in Sketch 7.1, not the entire sketch.

```
void MotorForward(){
 if (motorStep == 1) {
   digitalWrite(MotorPin1, MotorWireHigh);
   digitalWrite(MotorPin2, MotorWireLow);
   digitalWrite(MotorPin3, MotorWireLow);
   digitalWrite(MotorPin4, MotorWireHigh);
 }
 if (motorStep == 2) {
   digitalWrite(MotorPin1, MotorWireLow);
   digitalWrite(MotorPin2, MotorWireHigh);
   digitalWrite(MotorPin3, MotorWireLow);
   digitalWrite(MotorPin4, MotorWireHigh);
 }
 if (motorStep == 3) {
   digitalWrite(MotorPin1, MotorWireLow);
   digitalWrite(MotorPin2, MotorWireHigh);
   digitalWrite(MotorPin3, MotorWireHigh);
```

```
  digitalWrite(MotorPin4, MotorWireLow);
 }
 if (motorStep == 4) {
  digitalWrite(MotorPin1, MotorWireHigh);
  digitalWrite(MotorPin2, MotorWireLow);
  digitalWrite(MotorPin3, MotorWireHigh);
  digitalWrite(MotorPin4, MotorWireLow);
 }
 delay (MotorDelay);
 ++motorStep;
 if (motorStep > 4) {motorStep = 1;}
}

void MotorBackward (){
 if (motorStep == 4) {
  digitalWrite(MotorPin1, MotorWireHigh);
  digitalWrite(MotorPin2, MotorWireLow);
  digitalWrite(MotorPin3, MotorWireHigh);
  digitalWrite(MotorPin4, MotorWireLow);
 }
 if (motorStep == 3) {
  digitalWrite(MotorPin1, MotorWireLow);
  digitalWrite(MotorPin2, MotorWireHigh);
  digitalWrite(MotorPin3, MotorWireHigh);
  digitalWrite(MotorPin4, MotorWireLow);
 }
 if (motorStep == 2) {
  digitalWrite(MotorPin1, MotorWireLow);
  digitalWrite(MotorPin2, MotorWireHigh);
  digitalWrite(MotorPin3, MotorWireLow);
  digitalWrite(MotorPin4, MotorWireHigh);
 }
 if (motorStep == 1) {
  digitalWrite(MotorPin1, MotorWireHigh);
  digitalWrite(MotorPin2, MotorWireLow);
  digitalWrite(MotorPin3, MotorWireLow);
  digitalWrite(MotorPin4, MotorWireHigh);
 }
```

```
delay (MotorDelay);
--motorStep;
if (motorStep < 1) {motorStep = 4;}
}
```
Sketch 7.2

To use the motor as a half step system, change the MotorForward and MotorBackward subroutines in Sketch 7.1 to those in Sketch 7.3. Again, I will list only the subroutines that you need to replace in Sketch 7.1, not the entire sketch.

```
void MotorForward(){
if (motorStep == 1) {
digitalWrite(MotorPin4, MotorWireHigh);
digitalWrite(MotorPin3, MotorWireLow);
digitalWrite(MotorPin2, MotorWireLow);
digitalWrite(MotorPin1, MotorWireLow);
}
if (motorStep == 2) {
digitalWrite(MotorPin4, MotorWireHigh);
digitalWrite(MotorPin3, MotorWireLow);
digitalWrite(MotorPin2, MotorWireHigh);
digitalWrite(MotorPin1, MotorWireLow);
}
if (motorStep == 3) {
digitalWrite(MotorPin4, MotorWireLow);
digitalWrite(MotorPin3, MotorWireLow);
digitalWrite(MotorPin2, MotorWireHigh);
digitalWrite(MotorPin1, MotorWireLow);
}
if (motorStep == 4) {
digitalWrite(MotorPin4, MotorWireLow);
digitalWrite(MotorPin3, MotorWireHigh);
digitalWrite(MotorPin2, MotorWireHigh);
digitalWrite(MotorPin1, MotorWireLow);
}
if (motorStep == 5) {
```

```
digitalWrite(MotorPin4, MotorWireLow);
digitalWrite(MotorPin3, MotorWireHigh);
digitalWrite(MotorPin2, MotorWireLow);
digitalWrite(MotorPin1, MotorWireLow);
}
if (motorStep == 6) {
digitalWrite(MotorPin4, MotorWireLow);
digitalWrite(MotorPin3, MotorWireHigh);
digitalWrite(MotorPin2, MotorWireLow);
digitalWrite(MotorPin1, MotorWireHigh);
}
if (motorStep == 7) {
digitalWrite(MotorPin4, MotorWireLow);
digitalWrite(MotorPin3, MotorWireLow);
digitalWrite(MotorPin2, MotorWireLow);
digitalWrite(MotorPin1, MotorWireHigh);
}
if (motorStep == 8) {
digitalWrite(MotorPin4, MotorWireHigh);
digitalWrite(MotorPin3, MotorWireLow);
digitalWrite(MotorPin2, MotorWireLow);
digitalWrite(MotorPin1, MotorWireHigh);
}
delay(MotorDelay);
 ++motorStep;
 if (motorStep > 8) {motorStep = 1;}
}

void MotorBackward (){
if (motorStep == 8) {
digitalWrite(MotorPin1, MotorWireHigh);
digitalWrite(MotorPin2, MotorWireLow);
digitalWrite(MotorPin3, MotorWireLow);
digitalWrite(MotorPin4, MotorWireLow);
}
if (motorStep == 7) {
digitalWrite(MotorPin1, MotorWireHigh);
digitalWrite(MotorPin2, MotorWireLow);
```

```
digitalWrite(MotorPin3, MotorWireHigh);
digitalWrite(MotorPin4, MotorWireLow);
}
if (motorStep == 6) {
digitalWrite(MotorPin1, MotorWireLow);
digitalWrite(MotorPin2, MotorWireLow);
digitalWrite(MotorPin3, MotorWireHigh);
digitalWrite(MotorPin4, MotorWireLow);
}
if (motorStep == 5) {
digitalWrite(MotorPin1, MotorWireLow);
digitalWrite(MotorPin2, MotorWireHigh);
digitalWrite(MotorPin3, MotorWireHigh);
digitalWrite(MotorPin4, MotorWireLow);
}
if (motorStep == 4) {
digitalWrite(MotorPin1, MotorWireLow);
digitalWrite(MotorPin2, MotorWireHigh);
digitalWrite(MotorPin3, MotorWireLow);
digitalWrite(MotorPin4, MotorWireLow);
}
if (motorStep == 3) {
digitalWrite(MotorPin1, MotorWireLow);
digitalWrite(MotorPin2, MotorWireHigh);
digitalWrite(MotorPin3, MotorWireLow);
digitalWrite(MotorPin4, MotorWireHigh);
}
if (motorStep == 2) {
digitalWrite(MotorPin1, MotorWireLow);
digitalWrite(MotorPin2, MotorWireLow);
digitalWrite(MotorPin3, MotorWireLow);
digitalWrite(MotorPin4, MotorWireHigh);
}
if (motorStep == 1) {
digitalWrite(MotorPin1, MotorWireHigh);
digitalWrite(MotorPin2, MotorWireLow);
digitalWrite(MotorPin3, MotorWireLow);
digitalWrite(MotorPin4, MotorWireHigh);
```

```
}
 delay (MotorDelay);
 --motorStep;
 if (motorStep < 1) {motorStep = 8;}
}
```
Sketch 7.3

Since these are half steps, there are eight steps in a cycle, just as there were in Sketch 6.2. As with unipolar motors, you might also find that you can often use a shorter delay time (MotorDelay) operating in half step mode.

Chapter 8

Servo Motors

A rather unusual type of motor is the servo motor. Most motors rotate the shaft in the same direction over and over. Most servo motors rotate their shaft over a limited angle, usually 180 degrees. Figure 8.1 shows a popular servo motor for Arduinos, the SG 90 Tower Pro Micro Servo Motor. I will use this motor for my examples.

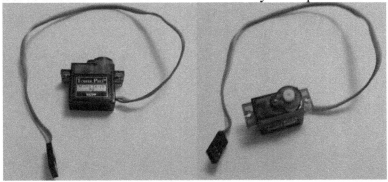

Figure 8.1

This motor has three wires with female connectors, so the easiest way to connect it to the Arduino is to use male-male wires (Dupont or 22-gauge single strand wire) to connect the female connectors on the Arduino to the female connectors on the Arduino. Connect the brown wire to GND and the orange wire to a digital output. It is best to use the Arduino pins 9 or 10, because the servo motor library that comes with the Arduino IDE automatically disables PWM on these pins anyway, so it is best to save other pins for other purposes. This motor can run on from 4.2 to 6 volts, so you can power it directly from the Arduino 5V output. However, you can also connect the red motor wire to the positive lead of an external power supply and the negative lead of the external power supply to one of the Arduino GND pins. This is generally a good idea, because I have found that under heavy loads, servo motors can draw so much power that they can cause the Arduino to reboot spontaneously. Also, connecting the servo to 6 volts, such as four AAA batteries, gives it more power. Figure 8.2 shows the schematic.

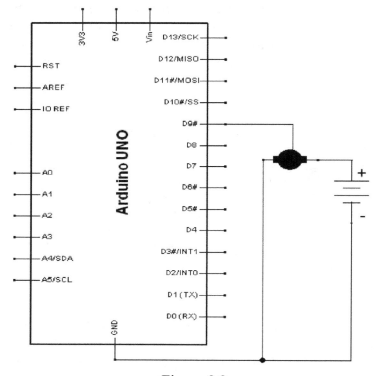

Figure 8.2

The next subject is the software. As mentioned above, the Arduino IDE comes with a library specifically designed to drive servo motors. Another unusual aspect of servo motors is the way that they are controlled. They are controlled by sending a short positive (HIGH) pulse to digital output pin the once every 20 milliseconds. The width of this pulse determines the position of the servo motor. Most commonly, a pulse of about 1000 microseconds sets the motor to the 0 degree position, a

2000 microsecond pulse sets it to 180 degrees, and pulses between 1000 and 2000 microseconds set it to positions proportional to the width. For example, 1250 microseconds sets it to 45 degrees, 1500 sets it to 90 degrees, etc. The good news is that you do not have to do all these calculations yourself. The servo library has functions that handle all this automatically. Sketch 8.1 demonstrates this by allowing you to send a setting in degrees from the serial monitor to the Arduino and have the Arduino set the servo to that number of degrees.

```
#include <Servo.h>
Servo myservo; // create servo object to control a servo
#define ServoPin 9
int pos = 90;

void setup() {
  myservo.attach(ServoPin);
  myservo.write(pos);
  Serial.begin(9600);
}

void loop() {
if (Serial.available() > 0) {
  pos = Serial.parseInt();
  Serial.print("pos = ");
  Serial.println(pos);
  myservo.write(pos);
  }
}
```
Sketch 8.1

The first line includes the Servo library in your sketch. This library comes built into the Arduino IDE, so you do not have to install it. The second line creates an object called myservo that will be used to address your servo motor to send instructions to it. The name does not have to be myservo. You can call it anything. In fact, you can create multiple servo objects. You simply define them with different names. For example, to define four servos, you could include the code

Servo myservo1;
Servo myservo2;
Servo myservo3;
Servo myservo4;

Again, these are just sample names. You can call each servo object whatever you want.

The next line defines the pin that you will connect to the servo motor. You need to set a different pin for each servo if you have more than one servo connected to the Arduino. The next line defines the variable, pos (short for position), that will hold the position you want the servo to be in. I have set it to 90 to start. It is not absolutely necessary to set the servo position at the beginning of the sketch, but it is best practice to start out your device with the servos in known positions. Note that setting pos to 90 does not immediately set the servo position. I am merely setting the variable for later use.

In the setup routine, I have first used the line
myservo.attach(ServoPin);
to assign the pin number that I had previously defined ServoPin as to the servo object myservo. If you have multiple servos, you can repeat this line with the different servo objects and pin numbers, such as
myservo1.attach(ServoPin1);
myservo2.attach(ServoPin2);
and so on. It might be worth noting that this is actually the only place where you will use the defined constant ServoPin. You could just as easily have simply inserted the number, such as 9 or 10, into the myservo.attach. However, defining all these constants at the beginning of the sketch makes it easy to find, and therefore change, what pins you have assigned to each servo.

The line
myservo.write(pos);

tells the servo motor to go to the position pos, which was set to 90. The positions are in degrees, so the servo will move to the 90 degree position. This is one of the two commands used to control the servo motors. The last line of the setup routine simply opens the communications to the serial monitor.

The loop routine simply accepts a number in degrees in the integer variable pos and uses the myservo.write command to send this to the motor. The myservo.write command converts the angle in degrees into a number of the proper number of microseconds for the pulse width.

The myservo.write command is based on the assumption that the servo motor you are using positions itself to a certain angle for a certain pulse width. Your servo motor might be a little different. It might range from 0 to 180 degrees, but use a different pulse width for a certain position. It might also have a different range of angles, like 0 to 90 degrees or 0 to 270. It might even have continuous rotation, which I will discuss shortly. This means that the angle of the motor might not correspond to the angle you tell it to set, which can be confusing. You can skip using the myservo.write command entirely and set the number of microseconds for the pulse width using the writeMicroseconds command. This uses the form
servoName.writeMicroseconds(number), where servoName is the name you gave your servo motor in the
Servo servoName;
Statement and number is the number of microseconds you want the pulse to be. Sketch 8.2 shows how to use direct microsecond numbers to control the servo motor.

#include <Servo.h>
Servo myservo; // create servo object to control a servo

#define ServoPin 9
int duration = 1500;

```
void setup() {
  myservo.attach(ServoPin);
  myservo.writeMicroseconds(duration);
  Serial.begin(9600);
}

void loop() {
  if (Serial.available() > 0) {
    duration = Serial.parseInt();
    Serial.print("Duration = ");
    Serial.println(duration);
    myservo.writeMicroseconds(duration);
  }
}
```
Sketch 8.2

This sketch actually works just like Sketch 8.1, except that I have substituted duration for pos (just a name change to reflect that it is a time duration instead of a position) and writeMicroseconds for write.

If you want to use some more convenient and intuitive measure than microseconds to express the position of the servo motor, such as degrees or even percent of rotation, but the servo motion does not match the degrees used by the built-in write function, you can use the Arduino map to convert. First, you can use Sketch 8.2 to determine what is the minimum and maximum number of microseconds your motor responds to. Just keep sending low numbers until the motor stops moving, and the lowest number that actually moved the motor is your low value for microseconds. Then increase the numbers until you reach the maximum number that causes a rotation, and record that number. Then choose what units you want to represent minimum and maximum values for you to input, such as 0 to 180 for degrees or 0 to 100 for percent. Sketch 8.3 shows how to use this information.

```
#include <Servo.h>
Servo myservo;  // create servo object to control a servo

#define ServoPin 9
#define minTime 600
#define maxTime 2400
#define minPosition 0
#define maxPosition 180


int duration;
int pos = 90;

void setup() {
  myservo.attach(ServoPin);
  duration = map(pos, minPosition, maxPosition, minTime, maxTime);
  myservo.writeMicroseconds(duration);
  Serial.begin(9600);
}

void loop() {
  if (Serial.available() > 0) {
    pos = Serial.parseInt();
    duration = map(pos, minPosition, maxPosition, minTime, maxTime);
    Serial.print("Duration = ");
    Serial.println(duration);
    Serial.print("position = ");
    Serial.println(pos);
    myservo.writeMicroseconds(duration);
  }
}
```

Sketch 8.3

At the beginning of your sketch, define the minimum time (minTime) and maximum (maxTime) for the pulse that produces rotation. Then define your units for

position. In this example, I used degrees and 0 to 180, but you can use any integer numbers. You could use 0 to 100 to indicate percent rotation. If the motor is being controlled by an input from an Arduino analog input, you can set these to the minimum and maximum possible input values you could receive, such as 0 to 1023. At any point in your sketch that you have a servo motor position in your units to convert to microseconds, use
duration = map(pos, minPosition, maxPosition, minTime, maxTime);
and then use
myservo.writeMicroseconds(duration);
to set the motor position.

As I mentioned earlier, there are some servo motors that are not limited to a set range of angles. These are called continuous rotation servo motors, and they spin in either direction depending on the control signal. They still use a pulse every 20 milliseconds. However, instead of telling the servo to go to a specific position, the pulse width tells it how fast to turn and in which direction. At a certain pulse width, usually about 1500 microseconds, the motor is stationary. For longer pulse widths, the motor turns in one direction. The longer the width, the faster it turns, up to a certain width (usually about 2000 or 2500 microseconds). For shorter pulse widths, the motor turns in the opposite direction. The shorter the pulse, down to about 1000 or 500 microseconds, the faster it turns.

The wiring is the same as for the other type of servo motor, so all the previous discussion about wiring and the schematic in Figure 8.2 apply and I will not repeat them. The wire colors might be slightly different, but generally the negative is black or brown, the positive red, and the input signal wire some other color such as white. Likewise, Sketches 8.2 and 8.3 can be used to control these motors. The best sketch to use is probably 8.3. You can call minPosition something like minSpeed and set it to something like -100, and call maxPosition something like maxSpeed and set it to 100. That way, inputting -100 will make it go maximum speed one direction, and inputting 100 will make it go maximum speed the other way.

Continuous rotation servo motors would be serving the same function as other types of motors, such as DC motors or stepper motors. They do not allow you as precise a control as stepper motors. The main advantage of continuous motion servo motors is that they are very small and light. This makes them very suitable for applications where you need a small, light motor such as in flying device like a drone or model plane, or you need a very small motor to go into a limited space, like a robot hand.

Chapter 9

Brushless Motors with ESC

Many types of motors, such as the typical DC motor, have strips of metal called brushes that slide over the rotating shaft to provide power to the electromagnets. These brushes can wear out, and are a major source of failure in motors. Brushless motors, at least the type I am talking about in this book, are motors that avoid using brushes by having an alternating current create changing magnetic fields that rotate a magnet attached to the shaft. Brushless motors of this type are particularly good for driving the shaft at very high speeds, such as you might need for a propeller on a drone or model airplane, a boat propeller, or even a fast model car. Unfortunately, they are rather complicated to operate because they require an alternating current (AC) instead of DC.

Because they require an alternating current to continuously shift the orientation of the magnetic fields, they cannot be driven by straight DC current. They require a circuit called an ESC (electronic speed control) to both convert the DC to AC and regulate the AC in the proper way to control the speed of the motor. Figure 9.1 shows two common ESC units.

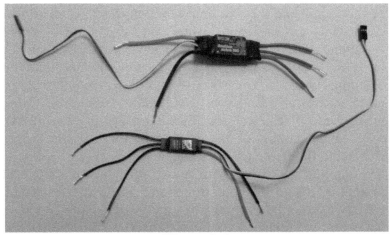

Figure 9.1

One side of the ESC has three wires coming from it, usually identical in appearance. These wires are arranged in a row. These are to be connected to the three wires on the brushless motor, which are also generally in a row. Figure 9.2 shows another ESC connected to the motor.

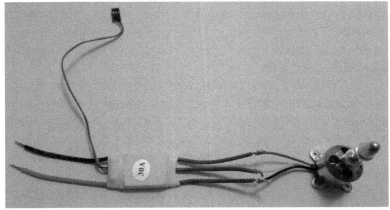

Figure 9.2

There are six possible combinations of connections between the controller and the motor. All will cause the motor to run. Some will cause it to run in one direction, some will cause it to run in the opposite direction. Some of the documentation I have read indicates that the middle wire from the ESC should be connected to the middle wire of the motor and the two end wires on the ESC should be connected to the end wires on the motor. Other documentation indicates that it does not matter. Although I have found that the motor does run under any configuration of connections, it does seem to run more smoothly if the middle wires are connected. I therefore recommend erring on the side of caution and always connecting the middle wire of the controller to the middle wire of the motor. Reversing the two end wires has no effect except reversing the motor direction. This can be useful if you want different motors to run in opposite directions. For example, if you have two motors on a model car, they will usually be facing opposite directions, so the motors need to run in opposite directions for both to drive the car forward. Therefore, connect motors on opposite sides of a car with the two end wires reversed. I will discuss other applications of reversing the leads later.

The two wires on the other side of the ESC connect to the power supply, with the red wire going to the positive on the power supply and the black wire going to the negative. The power supply generally needs to be able to provide a fairly high current, because brushless motors generally need a lot of power.

Between the two power wires are three smaller wires used to control the ESC. These are the wires you can connect to the Arduino. One of the end wires is usually black or brown, and goes to the Arduino GND. The middle wire is red. This can be used to power the Arduino by connecting it to the Arduino 5V, but I strongly advice against this. There are a quite a number of circumstances where using this to power your Arduino can damage your Arduino or even the computer you are using to program the Arduino, plus the fact that power spikes from this during motor operation can reboot the Arduino. I strongly recommend NOT connecting this red center wire to anything. The remaining wire, which is usually white or yellow but can vary, should be connected to an Arduino digital output, preferably 9 and/or 10 if you have one or two motors. I will explain why shortly.

You control the ESC from the Arduino in a manner very similar to the way you control a servo, as explained in Chapter 8. In fact, you even use the servo library. The speed of the brushless motor is controlled by the length of a pulse from the Arduino. Like the servo motor, the pulse should come every 20 milliseconds. Generally, the pulse width should be about 1 millisecond for the motor to be stopped and about 2 milliseconds for the motor to be going full speed, but you can change this. In fact, this is the tricky part.

You generally have to program the ESC to tell it what pulse width represents maximum speed and what pulse width represents minimum speed. The reason for this is that the ESC is usually connected to some type of controller instead of an Arduino. Different controllers may use different pulse widths for maximum and minimum speed, so the ESC allows you to "tell" the ESC what the pulse widths are when you connect it to a new controller. Once you have programmed the ESC with the desired pulse widths, you should not have to do it again unless you change the controller you are using. However, I have known the ESC to "forget" and need to be reprogrammed. I have found this to happen if the motor gets jammed, if you accidentally send it a pulse width higher or lower than the range you have set, or for other reasons.

Before discussing how to program the pulse range, you will need a sketch. Sketch 9.1 will allow you to control the ESC, and thus the motor.

```
#include <Servo.h>
Servo MYESC; // create ESC object to control an ESC

#define ESCPin 9
int duration = 1000;

void setup() {
```

```
  MYESC.attach(ESCPin);
  MYESC.writeMicroseconds(duration);
  Serial.begin(9600);
}

void loop() {
  if (Serial.available() > 0) {
    duration = Serial.parseInt();
    if (duration < 101) {duration=
map(duration,0,100,1000,2000);}
    Serial.print("Duration = ");
    Serial.println(duration);
    MYESC.writeMicroseconds(duration);
  }
}
```
Sketch 9.1

If you compare Sketch 9.1 with Sketch 8.2, you will see that they are really the same sketch, with some of the variable names and parameters changed. It even loads the servo library. That is because you are really doing exactly the same thing. You are sending a pulse of varying width every 20 milliseconds, using the servo library. I have changed the word servo to ESC to clarify what the sketch is used for, and I have changed the initial pulse width from 1500 to 1000 microseconds because I want the brushless motor begin at a speed of 0 while I wanted the servo to begin at 90 degrees, but otherwise it is the same sketch. As noted in Chapter 8, the servo library disables PWM on pins 9 and 10, which is why it is best to use those pins as your digital output, since they are partially disabled for other purposes anyway.

Now for the matter of how you program the pulse range that determines what pulse lengths indicate minimum stopped (stop) and maximum speed. In order to set, or reset, the pulse range, power up the Arduino with Sketch 9.1 loaded but do not connect battery power to the ESC. Set the Arduino pulse to the maximum width (such as 2000), which you want to represent top speed, by typing 2000 in the serial monitor. It should echo back "Duration = 2000." Then connect battery power to the ESC. The ESC should give off a series of tones, such as three beeps followed by two beeps. As soon as you hear the second set of beeps, input the low pulse width, such as 1000. You should do this very quickly after the second set of beeps (I will explain why shortly). Therefore, it is best to already have 1000 already typed into the serial monitor text box before you apply power to the ESC, but not press the ENTER key or click on the Send button. Then as soon as you hear the second set of beeps, press the ENTER key or click on the Send button to send the 1000. You should then hear another set of beeps from the ESC, such as three short beeps and a long one. The ESC is now programmed for pulse length 2000 to be the top speed signal and 1000 to be the stop signal. You can now tell the motor to go any speed between stopped and full speed by sending a number between 1000 and 2000. For example, 1100 will cause the motor to go fairly slowly, 1900 will cause it to go pretty fast, and 1500 will be mid range speed.

As mentioned before, the ESC should hold this information after you power down both the ESC and the Arduino. One it is programed in this way, always power up the Arduino first set to low pulse width (generally 1000), then apply power to the ESC. If the ESC finds that the input is the short pulses when you apply main power to it, it will not expect to be reprogrammed and will use the previously set minimum pulse length to indicate stop and the maximum pulse length to indicate full speed.

The reason that you need to send the 1000 very quickly after the second set of beeps is that the ESC is initially waiting to get the stop motor pulse width information. If it does not get that information within a few seconds, it will go into programming mode where it is expecting you to input other information, such as cutoff mode, cutoff threshold, etc. I will not go into these in this book because they vary with different ESC units and also because their applications depend on your use of the motor and are too complex to go into in this book. The point is that when programming the high and low pulse range to control speed, you should send the low pulse range signal (usually 1000) as soon as you hear the second set of beeps.

There are a number of ways you can improve the basic control given in Sketch 9.1. For one thing, brushless motors are very powerful and suddenly changing the rotation speed of the motor can cause tremendous strain on the motor mounts and related structure as well as whatever is connected to the motor (propeller, etc.). It also puts great strain on the battery. Causing the motor to speed up or slow down more gradually can alleviate some of these problems. Putting constraints on allowable settings can also be a useful safety feature. In addition, easier control commands like 0 to 100 (representing percent) can be more useful than trying to remember numbers like 1000 or 2000. Sketch 9.2 adds these improvements to Sketch 9.1.

```
#include <Servo.h>
Servo MYESC;  // create ESC object to control an ESC

#define ESCPin 9
#define maxDuration 2000
#define minDuration 1000

int duration = minDuration;
int newDuration = duration;
int stepSize = 5;
int slowSize = 100;
String S;

void setup() {
  MYESC.attach(ESCPin);
  MYESC.writeMicroseconds(duration);
  Serial.begin(9600);
}

void loop() {
```

```
if (Serial.available() > 0) {
  S = Serial.readString();
  S.toUpperCase();
  Serial.print(S);
  if (S == "L") {
    duration = minDuration;
    newDuration = duration;
    MYESC.writeMicroseconds(duration);
    S = "X";
  }
  if (S == "H") {
    duration = maxDuration;
    newDuration = duration;
    MYESC.writeMicroseconds(duration);
    S = "X";
  }
  if (S != "X") {
    newDuration = S.toInt();
    Serial.print("Input = ");
    Serial.println(newDuration);
    if (newDuration < 101) {newDuration= map(newDuration,0,100,minDuration,maxDuration);}
    if (newDuration < minDuration) {newDuration = minDuration;}
    if (newDuration > maxDuration) {newDuration = maxDuration;}
    if (newDuration > duration) {
      while (duration < newDuration){ //increase duration
        duration = duration + stepSize;
        Serial.print("Duration = ");
        Serial.println(duration);
        MYESC.writeMicroseconds(duration);
        delay(slowSize);
      }
      newDuration = duration;
    }
    if (newDuration < duration) { //decrease duration
      while (duration > newDuration){
```

```
      duration = duration - stepSize;
      Serial.print("Duration = ");
      Serial.println(duration);
      MYESC.writeMicroseconds(duration);
      delay(slowSize);
    }
    newDuration = duration;
   }
  }
 }
}
```

Sketch 9.2

I start off by defining maxDuration and minDuration since this sketch uses these values so often and this makes it easier to understand the sketch and also change these if necessary. I create the variable newDuration and set it to the current duration to allow code that gradually changes the current duration to the new setting. The variable stepSize is the size of the step you will change the rotation speed by, and slowSize is the variable that determines how long (in milliseconds) the code delays to allow the motor to adjust to each increment or decrement in speed. S is a string that will be used for input from the serial port.

In the main loop, the serial input inputs S. If S = "H" (for high), duration and newDuration are immediately set to the maximum value and this value is sent directly to the ESC. If S equals "L" (for low), duration and newDuration are immediately set to minimum value and this value is sent directly to the ESC. In either case, S is set to "X", which is just a flag to indicate that S is not a number. The purpose of this part of the code is to allow you to immediately set the pulse to the maximum and minimum values for programming the upper and lower range of pulse width, as explained previously. This is necessary because when you are programming the ESC, you must send these values immediately, not increment or decrement them slowly.

If S has not been set to "X," Meaning that is was not "L" or "H" when it was input, the line
newDuration = S.toInt();
concerts S to an integer and stores the value in newDuration, the pulse length duration you want to move to. If you input a value above 100, the code uses that value. However, if you input a value less than 101, it uses the map function to convert a number from 0 to 100 to a duration from 1000 to 2000. This allows you to input either a direct duration like 1500 for microseconds or a percent value like 50. The next two if statements make sure that the duration value is not less than minDuration or more than maxDuration.

Next, the process begins to gradually change the duration of the pulse width (and therefore the motor speed) from the current duration to the newDuration just input. If newDuration is greater than (>) duration (the current duration), the while loop slowly increases duration using the statement
duration = duration + stepSize;
The while loop continues to repeat this statement as long as duration < newDuration (duration is less than newDuration). The Serial.print commands just report the progress, and you can remove these. After each decrease in the variable duration, the
MYESC.writeMicroseconds(duration);
statement sends the new value of duration to the ESC, slowing the motor slightly. The delay command then pauses for the period slowSize to allow the motor to slow to match the new pulse duration. Note that you can change stepSize and slowSize to change the rate at which the motor adjusts to the new speed. Decreasing stepSize or increasing slowSize will slow down the rate of change of the motor speed.

Finally, the statement

newDuration = duration;
sets newDuration equal to duration. Normally, this would happen automatically since the entire process in the while loop was setting duration to the value of newDuration. However, it is possible for the values to not quite match at the end of the loop. For example, if stepSize is 10 and duration is 1000 and newDuration is 1105, the final duration will be 1110, since duration is increasing by 10. Having duration more than newDuration would cause problems when the code gets to the next part, which decreases duration if it is larger than newDuration. This would be a freak occurrence, but I am trying to make sure that no user mistakes can cause problems with this program. This does point out a small limitation of this sketch. You cannot set duration quite as precisely as you could in Sketch 9.2, since duration changes in steps of stepSize. Of course, you could always set stepSize to 1 and slowSize small, such as 10 or 20.

As mentioned above, the next set of code statements performs the opposite function, decreasing duration. As long as duration > newDuration, the statement
duration = duration - stepSize;
decreases duration by stepSize. It then sends the new duration to the ESC. As before, you can omit the Serial.print statements, although they can be useful during initial tests.

One technique not commonly covered with brushless motors is reversing them on the fly. That is, reversing them under programming control. There are situations where you might want to reverse the motors under programming control. For example, you might want to have a remote-controlled car or boat that can back up.

You may recall that I mentioned earlier in this chapter that if you reversed the wires to the motor, you can reverse the motor direction. You can reverse the motor wire connections easily with a relay. This is basically the same thing as we did in reversing a DC motor in Chapter 2 using a DPDT relay. However, brushless motors tend to be high current, so I recommend using a relay rated at at least 10 amps, and preferably 30. It is hard to find DPDT relays with ratings above one or two amps, so in Figure 9.3 I show the connections using two SPDT relays.

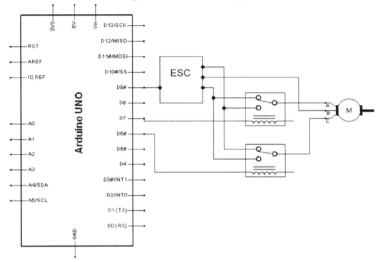

Figure 9.3

In this schematic I connect the relay control connections to Arduino outputs 6 and 7, although the outputs used do not matter. The center wire of the motor is connected to the center wire of the ESC. The wire on one end of the ESC is connected to the normally open contact of the first relay and the normally closed contact of the other. The other wire on one end of the ESC is connected to the normally closed contact of one the first relay and the normally open contact of the other. The center connection of each relay goes to one of the end connections of the motor. As usual in this book, I am not showing ground or power connections, but you understand that the power connections of the relays and the ground connection of the ESC must be connected to the Arduino. The code for this is shown in Sketch 9.3.

```cpp
#include <Servo.h>
Servo MYESC;  // create ESC object to control an ESC
#define ESCPin 9
#define relayOn LOW
#define relayOff HIGH
#define reversePin1 6
#define reversePin2 7

#define maxDuration 2000
#define minDuration 1000

int duration = minDuration;
int newDuration = duration;
int stepSize = 5;
int slowSize = 100;
String S;

void setup() {
  MYESC.attach(ESCPin);
  MYESC.writeMicroseconds(duration);
  pinMode(reversePin1, OUTPUT);
  pinMode(reversePin2, OUTPUT);
  digitalWrite(reversePin1, relayOff);
  digitalWrite(reversePin2, relayOff);
  Serial.begin(9600);
}

void loop() {
  if (Serial.available() > 0) {
    S = Serial.readString();
    S.toUpperCase();
    Serial.println(S);
    if (S == "L") {
      duration = minDuration;
      newDuration = duration;
      MYESC.writeMicroseconds(duration);
      S = "X";
    }
```

```
if (S == "H") {
  duration = maxDuration;
  newDuration = duration;
  MYESC.writeMicroseconds(duration);
  S = "X";
}
if (S == "F") {
  if (duration != minDuration) {
    duration = minDuration;
    newDuration = duration;
    MYESC.writeMicroseconds(duration);
    delay(500);
  }
  digitalWrite(reversePin1, relayOff);
  digitalWrite(reversePin2, relayOff);
  S = "X";
}
if (S == "R") {
  if (duration != minDuration) {
    duration = minDuration;
    newDuration = duration;
    MYESC.writeMicroseconds(duration);
    delay(500);
  }
  digitalWrite(reversePin1, relayOn);
  digitalWrite(reversePin2, relayOn);
  S = "X";
}
if (S != "X") {
  newDuration = S.toInt();
  Serial.print("Input = ");
  Serial.println(newDuration);
  if (newDuration < 101) {newDuration= map(newDuration,0,100,minDuration,maxDuration);}
  if (newDuration < minDuration) {newDuration = minDuration;}
  if (newDuration > maxDuration) {newDuration = maxDuration;}
```

```
    if (newDuration > duration) {
      while (duration < newDuration){ //increase duration
        duration = duration + stepSize;
        Serial.print("Duration = ");
        Serial.println(duration);
        MYESC.writeMicroseconds(duration);
        delay(slowSize);
      }
      newDuration = duration;
    }
    if (newDuration < duration) { //decrease duration
      while (duration > newDuration){
        duration = duration - stepSize;
        Serial.print("Duration = ");
        Serial.println(duration);
        MYESC.writeMicroseconds(duration);
        delay(slowSize);
      }
      newDuration = duration;
    }
   }
  }
 }
}
```

Sketch 9.3

This is similar to Sketch 9.2, with some extra code added to control the relay. The relay pins and relayOn and relayOff are defined, much like in Chapter 2. The relay pins are initialized for OUTPUT and set to off in the setup routine. In the main loop, I have added code to respond to inputting "F" (for forward) and "R" (for reverse) from the Serial monitor. When you input either of these commands, the code first checks to see if the motor is currently running (duration is not minDuration). It could be rather bad to suddenly switch direction if the motor is running. If the speed is not zero, duration is set to minDuration and newDuration is set to the new duration. The new duration is then sent to the ESC with the

MYESC.writeMicroseconds(duration);
Statement. The code then delays for 5 milliseconds to give the motor time to stop. You can adjust this time if you like. After the motor has stopped (if that was necessary), the relays are both turned on if you sent the "R" command and Off is you sent the "F" command. S is then set to "X" to make sure the speed control portion of the code is not run. Other than these changes, it is the same as Sketch 9.2

Chapter 10

Additional Information

If you want to control the motors by remote control, I recommend that you purchase another one of my books, "Remote Sensor Monitoring by Radio with Arduino: Detecting Intruders, Fires, Flammable and Toxic Gases, and Other Hazards at a Distance." This book is available on Kindle or paperback from Amazon.com, and contains considerable information about using the nRF24L01 transceiver chip to send information between two or more Arduinos. Although the book is about connecting sensors to this chip and transmitting the sensor data, you can easily replace the sensors with a joystick, transmit the readings to another Arduino, and use the numbers provided in the same way you used the input from the serial monitor in the sketches in this book. The data is generally transmitted as integer arrays, and you can use the values to control speed, direction, or other aspects of the motor just as you used the integer values that you got from the serial port in this book.

If you use input from sensors connected to your Arduino to control the speed of a motor, you need to convert the values from the sensor to the appropriate values to control the motor. For example, if you are controlling a servo of brushless motor, you need to convert the sensor reading to a pulse width from about 1,000 microseconds to 2,000 microseconds. If you are controlling the speed of a DC motor, you need to convert the sensor readings to a PWM value from 0 to 254. In Sketch 6.3, the speed of the controller motor was controlled by the variable control. In any case, you will probably use the map function to convert the input value to the appropriate motor speed. The format, as I have shown in some of the sketches in this book, is motorControlVariable = map(inputVariable,MinInput, MaxInput,MinMotorValue,MaxMotorValue);

When you do this, it is important to know the minimum and maximum range limit of the input values will be, as well as the minimum and maximum desired motor variables are. In the examples in this book, I generally had a minimum input of 0 and a maximum input of 100, but that was just because the numbers were being input by a person. Your real-world input values can have a minimum and maximum of various values, so be sure to set your values in your map function accordingly.

You can download the sketches in this from https://github.com/DavidLeithauser/Files-for-Controlling-Motors-with-Arduino-Genuino-

They can also be downloaded at this time from https://LeithauserResearch.com/MotorsBook.zip

This is my business Web site, and will be active until at least February 2020, possibly longer. The sketches in this ZIP file are in a folder called MotorsBook. The sketches are in the format S#P#, which stands for Sketch # Point #. For example, Sketch 2.1 is called S2P2.ino in the archive. If you download the ZIP file and extract it, then move the MotorsBook folder to the Arduino folder in your computer's documents folder, the MotorsBook folder and its sub folders should appear in Sketchbook under the File menu at the top left of the Arduino IDE.

You can contact me at Leithauser@aol.com if you have any questions, comments, problem reports, etc. about the book. Of course, while I like helping people by providing technical help with my books, I cannot guarantee how much time I can devote to helping any one person at any given time.

CPSIA information can be obtained
at www.ICGtesting.com
Printed in the USA
LVHW080455030521
686321LV00024B/590